Continuity and Change

ALEXANDER PAPAGEORGIOU

Preservation in City Planning

with a Preface by Frederick Gutheim

PRAEGER PUBLISHERS
New York · Washington · London

Translated by Gerald Onn

Praeger Publishers, Inc.
111 Fourth Avenue, New York, N.Y. 10003, U.S.A.
5 Cromwell Place, London, S.W.7, England

Published in the United States of America in 1971
Originally published as *Stadtkerne im Konflikt*
© 1970 by Verlag Ernst Wasmuth, Tübingen
Translation © 1971 by Pall Mall Press Limited, London, England

Printed in Germany

IN MEMORY OF MY BELOVED
AND REVERED TEACHER,
DEMETRIUS PIKIONIS,
GREEK ARCHITECT AND AESTHETE

CONTENTS

Illustrations

Bibliography

A: General History of Art and Architecture. Archaeology

ΒΑΣΙΛΕΙΑΔΗ ΔΗΜΗΤΡΙΟΥ: Εἰσαγωγὴ εἰς τὴν αἰγαιοπελαγίτικη ἀρχιτεκτονική. ᾿Αθῆναι 1955

ΒΑΣΙΛΕΙΑΔΗ ΔΗΜΗΤΡΙΟΥ: ῾Η λαϊκή ἀρχιτεκτονικὴ τῆς Αἴγινας. ᾿Αθῆναι 1957–8

ΒΑΣΙΛΕΙΑΔΗ ΔΗΜΗΤΡΙΟΥ: Τὸ ἀχειροποίητο σύμπλεγμα τῆς παραπορτιανῆς Μυκόνου. ᾿Ανάτυπον, τεῦχος 27 «᾿Αρχιτεκτονική». ᾿Αθῆναι 1961

BENOIS, FERNAND: *Arles.* Alpina, Paris 1959

CHADZIDAKIS, MANOLIS: *Das Byzantinische Athen.* M. Pechlivanides, Athens

CHIERICHETTI: *Florence.* Innocenti, Florence

DIEHL, CHARLES: *Salonique.* Henri Laurens, Paris 1920

DJELEPY, PANOS: *L'architecture populaire en Grèce.* Morancé, Paris 1952

EPPENS, HANS: *Baukultur im Alten Basel.* Frobenius, Basel 1953

ΚΩΝΣΤΑΝΤΙΝΙΔΗ ΑΡΗ: Δύο χωριὰ ἀπὸ τὴ Μύκονο. ᾿Αθῆναι 1948

LACARRIERE, JACQUES: *Mont Athos.* Seghers, Paris 1954

LAMPAKIS, GEORGES: *Mémoire sur les Antiquités Chrétiennes de la Grèce.* Hestia, Athens 1902

LEHMANN, FRIEDRICH: *Kurfürstendam.* Herbig, Berlin 1964

ΜΑΘΙΟΠΟΥΛΟΥ ΠΑΝ.: ῞Αγιον ὄρος. «᾿Αρχιτεκτονική» ἀρ. 48. ᾿Αθῆναι Νοεμ.-Δεκεμ. 1964

ΜΑΘΙΟΠΟΥΛΟΥ ΠΑΝ.: ῾Η Μακρυνίτσα καὶ ἡ λαϊκὴ τέχνη τοῦ Πηλίου. «᾿Αρχιτεκτονική» ἀρ. 51. ᾿Αθῆναι Μάϊος-᾿Ιουν. 1963

ΜΑΡΑΒΑ-ΧΑΤΖΗΝΙΚΟΛΑΟΥ, ANNA: Πάτμος. Institut Français d'Athènes. ᾿Αθῆναι 1957

MATTON, RAYMOND: *Athènes et ses monuments.* Institut Français d'Athènes, Athens 1963

ΜΟΥΤΣΟΠΟΥΛΟΣ ΝΙΚ.: Καστοριά, τὰ ἀρχοντικά. ῎Εκδοσις «᾿Αρχιτεκτονικῆς». ᾿Αθῆναι 1962

ΜΠΙΡΗ ΚΩΝΣΤΑΝΤΙΝΟΥ: Αἱ ᾿Αθῆναι ἀπὸ τοῦ 19ου ἕως τὸν 20ον αἰῶνα. Καθίδρυμα πολεοδομίας καὶ ἱστορίας τῶν ᾿Αθηνῶν. ᾿Αθῆναι 1966

MÜNTER, GEORG: *Idealstädte.* Henschel Verlag, Berlin 1957

MYLONAS, PAUL: *L'architecture du Mont Athos.* Institut Grec de Venise, Venice 1963

PAPAHATIJS, NICOLAS: *Monuments of Salonica.* S. Molho, Salonica 1962

PEVSNER, NIKOLAUS: *An Outline of European Architecture.* Penguin Books, London 1951

SHERRARD, PHILIP: *Athos, der Berg des Schweigens.* Urs Graf, Lausanne

SMITH, ARNOLD: *The Architecture of Chios.* Alex Tiranti, London 1962.

ΦΡΟΝΤΙΣΤΗΡΙΑΚΑΙ ΕΡΓΑΣΙΑΙ ΕΔΡΑΣ ΜΟΡΦΟΛΟΓΙΑΣ-ΡΥΘΜΟΛΟΓΙΑΣ Ε.Μ.Π.: Τὸ ῾Ελληνικὸ λαϊκό Σπίτι. ᾿Εθνικὸν μετσόβιον πολυτεχνεῖον. ᾿Αθῆναι 1960

VÖGELIN, S.: *Das alte Zürich.* Zürich, Vol. 1 1878. Vol. 2 1890

BABELON-FLEURY-DE SACY: *Richesse d'art du quartier des Halles.* Flammarion, Paris 1967

᾿Αρχοντικὰ Καστοριᾶς. Σύλλογος ῾Ελληνικὴ λαϊκὴ τέχνη. ᾿Αθῆναι 1948

Corinth. American School of Classical Studies, Athens 1964

Deutschland: Dome und Kirchen. Deutsche Zentrale für Fremdenverkehr, Hanover 1953

Guide des Parcs et Châteaux de France. Commissariat Général au Tourisme, Paris 1960

Mykonos. Bericht einer Studentenexkursion unter Pr. Hermkes und Pr. Hammerbacher. Technical University, Berlin, Berlin 1966

Παληά ᾿Αθήνα. ῎Ασυλον τέχνης. ᾿Αθῆναι 1931

Σπίτια τῆς Ζαγορᾶς. Σύλλογος ῾Ελληνικὴ λαϊκὴ τέχνη. ᾿Αθῆναι 1948

The Middle Ages in the Athenian Agora. American School of Classical Studies, Princeton 1961

B: Aesthetics and the Theory of Architecture

ABEL, ADOLF: *Vom Wesen des Raumes in der Baukunst.* Callwey, Munich 1952

ΒΑΣΙΛΕΙΑΔΗ ΔΗΜΗΤΡΙΟΥ: ῾Η διαμόρφωσις τῶν λόφων γύρω ἀπὸ τὴν ᾿Ακρόπολι. Τὸ ἔργο τοῦ Δημ. Πικιώνη. «᾿Αρχιτεκτονική» ἀρ. 36. ᾿Αθῆναι Νοεμ.-Δεκεμ. 1963

CONRADS-SPERLICH: *Phant. Archit.* Hatje, Stuttgart 1959

CORBUSIER, LE: *Vers une architecture.* Vincent Fréal, Paris 1920, 1958

Different Articles in *Architectura CSSR:* Gothic Cities in Czechoslovakia

Notes on the Architecture of the Renaissance in Slovakia.

Baroque Architecture in Bohemia and Moravia.

Popular Architecture in Czechoslovakia. No. 9–10 1966

FISCHER, THEODOR: *Zwei Vorträge zur Proportion.* Oldenburg, Munich–Berlin 1934

FRIEDMAN, YONA: *Mobile Architecture,* Paris 1963

JOUVEN, GEORGES: *Rythme et architecture.* Vincent Fréal, Paris 1951

ΚΩΝΣΤΑΝΤΙΝΙΔΗ ΔΗΜ.: Περὶ ἁρμονικῶν χαράξεων εἰς τὴν ἀρχιτεκτονικὴν καὶ τὰς καλὰς τέχνας. ᾿Αθῆναι 1961

ΜΙΧΕΛΗ ΠΑΝΑΓ.: Αἰσθητικὴ θεώρησις τῆς Βυζαντινῆς τέχνης. ᾿Αθῆναι

ΜΙΧΕΛΗ ΠΑΝΑΓ.: Αἰσθητικὰ θεωρήματα Α. ᾿Αθῆναι 1962, Β. ᾿Αθῆναι 1965

ΜΙΧΕΛΗ ΠΑΝΑΓ: ῾Η αἰσθητικὴ τῆς ὀπτικῆς ἀπάτης στὴν ἀρχιτεκτονική. ᾿Αθῆναι 1951

ΜΙΧΕΛΗ ΠΑΝΑΓ: ῾Η ἀρχιτεκτονικὴ ὡς τέχνη. ᾿Αθῆναι 1965

PEREITZ, ALEXANDRE: Vers un urbanisme spatial. *Architecture d'aujourd'hui*, No. 101, 1962, pp. 90–9

RUDOFSKY, BERNARD: *Architecture without Architects*. The Museum of Modern Art, New York 1965

SCHUBERT, OTTO: *Optik in Architektur und Städtebau*. Gebrüder Mann, Berlin 1965

SHARP, THOMAS: *Town and Townscape*. John Murray, London

SIEDLER-NIGGEMEYER-ANGRESS: *Die gemordete Stadt*. Herbig, Berlin 1964

SPOERRI, THEOPHIL: *Der Weg zur Form*. Furche, Hamburg 1954

UTUDJIAN, EDOUARD: *Architecture et Urbanisme Souterrains*. Robert Laffont, Paris 1966

WIESER, W.: *Organismen, Strukturen, Maschinen*. Frankfurt/Main 1959

ZUCKER, PAUL: *Entwicklung des Stadtbildes*. Drei Masken Verlag, Munich–Berlin 1929

Panoptikum: Architektur und Fotografie. Camera, Lucerne. No. 12, Dec. 1967

Περιοδικόν "Ζυγός" Τεῦχος ἀφιερωμένο στό ἔργο τοῦ Δημ. Πικιώνη. ἀρ. 27–28. 'Αθῆναι 1958

C: The Rehabilitation of Historic Settlements and Architectural Monuments. Theory and Strategy of Development

BANDINELLI, RANUNCIO: *La reconstruction artistique en Italie*. Istituto poligrafico dello stato, Rome 1947

BAUD, BOVY: Τὰ Σπίτια τοῦ Καλέσοβο εἰς τὴν "Ηπειρο («'Αρχιτεκτονική» ἀρ. 39. 'Αθῆναι Μάϊος–'Ιουν. 1963

BORCHERDT, HELMUT: Neue Fußgängerbereiche in der Münchener Altstadt. *Bauwelt* No. 37, Berlin Sept. 1966

CLEMENT, PAUL: *Die Deutsche Kunst und Denkmalpflege*. Kunstverlag, Berlin 1933

DE CARLO, GIANCARLO: *Urbino*. Marsilio Editore, 1966

DEMORIANE, HELENE: Ainsi fut Rome. *Connaissance des arts*, No. 146, Paris Apr. 1964

ERNAUD, FRANÇOIS: Le mont Saint Michel à la recherche de sa dureté. *Connaissance des arts*. No. 153, Paris Nov. 1964

ΖΗΒΑ ΔΙΟΝΥΣΙΟΥ: Τὸ ἔργο τῆς 'Ελληνικῆς ἀρχαιολογικῆς ὑπηρεσίας. 'Ανάτυπο τεχνικῶν χρονικῶν/τεύχ. 2. 'Αθῆναι Μάρτιος 1965

GOEDERITZ, JOHANNES: *Stadterneuerung*. Bauverlag, Wiesbaden–Berlin 1962

GOEDERITZ, JOHANNES: *Sanierung erneuerungsbedürftiger Baugebiete*. Krämer, Stuttgart 1960

GORMSEN, NIELS: Altstadtsanierung Bietigheim. *Bauwelt* No. 39–40, Berlin Sept. 1966

GRAUTOFF, OTTO: *Kunstverwaltung in Frankreich und Deutschland*. Max Drechsel, Berne 1915

HORLER, MIKLOS: Die Wiederherstellung des Salomonsturmes in Visegrad, Ungarn. *Bauwelt No. 30*, Berlin July 1968.

KUEHNE, GÜNTHER: Das Haus der Bürgerschaft in Bremen. *Bauwelt No. 6–7*, Berlin Feb. 1967

LAVANIGNO, EMILIO: *Fifty war damaged monuments of Italy*. Istituto poligrafico dello stato, Rome 1946

MICHAELIDIS, CONSTANTIN: *Hydra: A Greek island town, its growth and form*. The University of Chicago Press. Chicago and London 1969

MICHELUCCI, GIOVANNI: *Il quartiere di Santa Croce nel futuro di Firenze*. Officina Edizioni, Florence 1967

MINISTRY OF HOUSING AND LOCAL GOVERNMENT: *Historic towns. Preservation and change*. Her Majesty's Stationery Office, London 1967

ΜΟΥΤΣΟΠΟΥΛΟΥ ΝΙΚ.: Θεσσαλονίκη. 'Αρχιτεκτονικὰ μνημεῖα καὶ προβλήματα. («'Αρχιτεκτονική» ἀρ. 35. 'Αθῆναι Σεπτ.–'Οκτ. 1962

ΠΑΠΑΓΕΩΡΓΙΟΥ ΑΛΕΞ.: Πλάκα. Μία πρόταση γιὰ τὴν παλιά πόλι. 'Αθῆναι 1965

PELLATI, FRANCESCO: Résurrection des monuments en Italie. Article dans le numéro spécial "Italie" de la revue 'l'Amour de l'Art'

PLATH, HELMUT: Hannover, Historisches Museum am Hohen Ufer. *Bauwelt No. 9*, Berlin Feb. 1967

PUCCI, EUGENIO: *The flood in Florence*. Bonnechi, Florence 1966

RAUDA, WOLFGANG and WURZER, RUDOLF: *Salzburg*. Patzer Verlag, Hanover–Berlin 1968

REVEL, JEAN-FRANÇOIS: Plaidoyer pour Viollet-le-Duc. *Connaissance des arts*, Paris Sept. 1960

RICHEZ, ROGER: *Etude opérationnelle de restauration du secteur sauvegardé de la ville de Bourges*. Paris 1967

SELDMAYER, HANS: *Die demolierte Schönheit*. Otto Müller-Verlag, Salzburg 1965

SCIMEMI, GABRIELE: Venedig ist gerettet. *Bauwelt No. 29*, Berlin July 1964

Städtebauliches Seminar 'Regensburg', unter der Leitung Prof. W. HEBBEBRAND: *Regensburg. Zur Erneuerung einer alten Stadt*. Econ Verlag, Düsseldorf–Vienna 1967

STEIN, RUDOLF: *Das Sparkassenhaus am Markt zu Bremen*. Bremen 1968

ΤΡΑΓΛΟΥ ΙΩΑΝΝΟΥ: 'Ανασκαπτέος χῶρος τῶν 'Αθηνῶν, βασικὸν ζήτημα τῆς πολεοδομίας των. («'Αρχιτεκτονική» ἀρ. 38. 'Αθῆναι Μάρτ.–'Απρ. 1963

WARD, PAMELA: *Conservation and Development in Historic Towns and Cities*. Oriel Press, Newcastle upon Tyne 1967

WITMER, JOHN: Historische Stadtviertel. *Schweizerische Bauzeitung*. No. 18, pp. 313–9, May 1967

WORSKETT, ROY: *The Character of Towns, an Approach to Conservation*. The Architectural Press, London 1969

ZACHATOWITCZ, JAN: *La protection des monuments historiques en Pologne*. Polonia, Warsaw 1965

Alt und Neu. Wanderausstellung des B.D.A. *Sonderdruck der Deutschen Bauzeitung*, Stuttgart 1958

A ne pas démolir. *Connaissance des arts*. No. 144, Paris Feb. 1964

Bewahren und Gestalten. Deutsche Denkmalpflege. Von Holten, Berlin 1965

Coventry Rebuilds. Special issue of *Architectural Design*, Dec. 1958

Les Trésors Secrets de Beaune. *Connaissance des arts*. No. 150, Paris August 1964

Numéro spécial sur l'éclairage des monuments. *Revue internationale de l'éclairage*, No. 6/1965

Renaissance de Royaumont. *Connaissance des arts*, Paris Sept. 1960

Paris. *L'architecture d'aujourd'hui* No. 138, Paris June July 1968

Sauvegarde et mise en valeur du Paris historique. *Festival du Marais*, Paris 1967

Symposium 1964. Die Sanierung der Spandauer Altstadt. Technical University Berlin, Berlin 1964

The Stoa of Attalos II in Athens. American School of Classical Studies, Princeton

Umbauten, Renovierungen, Erweiterungen, Anbauten, Aufstockungen. *Baumeister No. 9*, Munich Sept. 1968

D: The Rehabilitation of Historic Settlements and Architectural Monuments. Legislation and Administrative Measures

Edited by Dr. H. JUNGMANN: *Gesetz zum Schutze von Kunst-, Kultur- und Naturdenkmälern*. Kupky, Dresden 1934

HEYER, KARL: *Denkmalpflege und Heimatschutz im Deutschen Recht*. Heymanns, Berlin 1912

Kultusminister des Landes Schleswig-Holstein: *Gesetz zum Schutze der Kulturdenkmäler des Landes*. Nachrichtenblatt des Kultusministers des Landes Schleswig-Holstein No. 1, Kiel 1961

WITMER, JOHN: *Unterlagen für die Bearbeitung des Altstadt-Baureglements der Stadt Zug*. 1967

Loi No. 62902 du 4 Août 1962 sur la protection du patrimoine historique et esthétique. *Journal Officiel de la République Française*, Apr. 1962

Decret No.63691 du 13 Juillet 1963 sur la protection du patrimoine historique et esthétique. *Journal Officiel de la République Française* July 1963,

Νόμος 1469 τοῦ 1950 τοῦ βασιλείου τῆς Ἑλλάδος. Ἀθῆναι 1950

Salzburger Altstadterhaltungsgesetz (In '*Salzburg*' von W. Rauda und R. Wurzer). Patzer-Verlag, Hanover–Berlin 1968

E: Town Planning and the History of Town Planning. Regional Planning

ABEL, ADOLF: *Die Regeneration der Städte*. Zürich 1950

ARAVANTINOS, A.: *Großstädtische Einkaufszentren.* Vulcan, Essen 1963

BRUNNER, CARL: *Städtebau und Schnellverkehr*. Springer, Vienna 1960

CORBUSIER, LE: *Grundfragen des Städtebaues*. Gerd Hatje, Stuttgart 1945

CORBUSIER LE: *Urbanisme*. Vincent Fréal, Paris 1924

ΔΕΣΠΟΤΟΠΟΥΛΟΥ Ι.: Προβλήματα, νεώτερες κατευθύνσεις στήν σύγχρονη πολεοδομία. «Ἀρχιτεκτονική» ἀρ. 33. Ἀθῆναι Μαϊος/Ἰουν. 1962

ΔΟΞΙΑΔΗΣ Κ.: Ἡ πρωτεύουσα μας καί τό μέλλον της. Τέχν. γραφεῖον Δοξιάδη. Ἀθῆναι 1960

DOXIADIS, CONSTANTIN: *Raumordnung im Griechischen Städtebau*. Vowinckel, Heidelberg–Berlin 1937

EGLI, ERNST: *Die neue Stadt in Landschaft und Klima*. Verlag für Architektur, Zürich 1951

EGLI, ERNST: *Geschichte des Städtebaues*. Eugen Rentsch, Erlenbach-Zürich-Stuttgart 1959–67

ΖΕΝΕΤΟΥ ΤΑΚΗ: Ἡ πόλι καί ἡ κατοικία στό μέλλον. «Ἀρχιτεκτονική» ἀρ. 42. Ἀθῆναι Νοεμ./Δεκεμ. 1963

GERKAN, ARNIM VON: *Griechische Städteanlagen*. De Gruyter, Berlin–Leipzig 1924

GIULIANO, ANTONIO: *Urbanistica delle città greche*. Il Saggiatore, Milan 1966

GRUEN, VICTOR: *The Heart of Our Cities*. Simon & Schuster, New York 1964

HARTOG, RUDOLPH: *Stadterweiterungen im 19. Jahrhundert*. Kohlhammer, Stuttgart 1962

HILBERSHEIMER, L.: *The Nature of Cities*. Paul Theobald & Co., Chicago 1955

JASPERT, FRITZ: *Vom Städtebau der Welt*. Safari, Berlin 1961

JONAS, WALTER: *Das intra Haus*. Origo, Zürich 1962

KIKUTAKE-KAWAZOE-OTAKA-MAKI-KUROKAWA: *Metabolism: The Proposals for a New Urbanism*. Japan 1960

KRAYENBUEHL, FRANK: *Untersuchung über die Entstehung und das Wachstum der Zentren in der Stadt Zürich*. Juris, Zürich 1963

KRIESIS, ANTHONY: *Greek Town Building*. National Techn. University of Athens, Athens 1965

LAVEDAN, PIERRE: *Les Villes Françaises*. Vincent Fréal, Paris 1960

MARTIN, ROLLAND: *L'urbanisme dans la Grèce Antique*. Picard, Paris 1956

PAPAGEORGIOU, ALEXANDER: *Athènes Majeure*. Vincent Fréal, Paris 1970

PAPAS, CONSTANTIN: *L'urbanisme et l'architecture populaire dans les Cyclades*. Dunod, Paris 1957

RAGON, MICHEL: *Où vivrons nous demain?* Robert Laffont, Paris 1963

RAGON, MICHEL and others: *Les Visionnaires de l' Architecture*. Robert Laffont, Paris 1965

RAUDA, WOLFGANG: *Raumprobleme im europäischen Städtebau*. Callwey, Munich 1956

RAUDA, WOLFGANG: *Lebendige städtebauliche Raumbildung*. Julius Hoffmann-Verlag, Stuttgart 1957

RESTANT, PIERRE: Πρὸς μία φανταστική ἀρχιτεκτονική. «Ἀρχιτεκτονική» ἀρ. 47. Ἀθῆναι Σεπτ./Ὀκτ. 1964

ROBINSON, HENRY: *The Urban Development of Ancient Corinth*. American School of Classical Studies, Athens 1965

SCHNEIDER, W.: *Überall ist Babylon. Die Stadt als Schicksal des Menschen von Ur bis Utopia*. Düsseldorf 1960

SIMON, HANS: *Das Herz unserer Städte*. R. Bacht, Essen. Vols. I and II, 1963–5

SITTE, CAMILLO: *Der Städtebau nach seinen künstlerischen Grundsätzen*. Vienna 1889

SPECKTER: *Städtebauliche Entwicklung von Paris*. Callwey, Munich

ΤΣΑΓΚΑΡΙΔΗ Ο.: Ἡ σύγχρονος ὄψις τῆς παλαιᾶς καί νέας πόλεως τῶν Ἀθηνῶν (Ἀεροφωτ.) «Ἀρχιτεκτονική» ἀρ. 21. Ἀθῆναι Μάϊος/Ἰουν. 1960

WYCHERLEY: *How the Greeks build Cities*. Macmillan, London 1949

Destin de Paris. *Urbanisme No.84*, Paris 1964

Städtebau. *Atlantis No.3*, Zürich March 1962

F: Works of general interest

EGLI, EMIL: *Das Flugbild Europas*. Artemis, Zürich 1958

FOUGERES, GUSTAVE: *Paris vu du ciel*. Armand Colin, Paris 1956

HENRARD, ROGER: *Athènes*. Laurens, Paris 1914

PREFACE by Frederick Gutheim

The historic centres of cities have been a casualty of contemporary urban growth and change. Population pressures, changing metropolitan structure, greatly increased mobility, an affluence having little to do with the quality of urban life – all these have had their impact on the traditional centres of cities. Their effect is seen in urban redevelopment, in efforts to accommodate increased traffic and to provide parking, in suburban growth and urban decay. Many city centres have been obliterated by such changes. Others have lost their traditional functions as centres of retail trade, finance, culture, education and even government. While they may retain some reassuring aspect of their former appearance, they are really living on their past momentum or awaiting some future change. Great metropolitan centres, like New York City, have totally lost the great buildings and urban places associated with their history and, as in Lower and Midtown Manhattan, are speedily losing the lesser ones. The historic centres that come easily to mind, like Annapolis, really owe their preservation to the accidents of growth that have protected them against change rather than (until very recent times) deliberate community action; and their future survival is far from assured.

Today there is greater awareness of the cultural loss in such mindless destruction. Indeed, there is a mood of apprehension. But slowly the tools of historic area preservation are being forged. And since the immense task cannot be accomplished at once, it must be decided what historic areas must be saved first, what their boundaries should be, what activities they should contain, what measures of change are compatible with the inheritance of the past. Humanistic values require clear definitions, and historic area preservation must create a strong theoretical framework in order to answer the flood of practical issues it faces in many cities.

It is to provide such a foundation for the preservation of historic areas that the present book has been written. It is the first to reflect extensive experience in many countries, and to relate this experience to a consistent set of ideas and a body of practice. Before facing the specifics of individual historic area design problems, or even the substantial questions of legal powers and administrative frameworks, the concept of the cultural monument must be stated. Only thus can the goals and objectives of such activity rest upon a solid foundation in human values, providing the basis for political action and, still more fundamentally, of public agreement. The need for such a statement has appeared only in the last decade. It is related not only to the increasing awareness of the emperilled historic centres of cities, but to the emergence of new urban planning techniques and the appreciation of environmental quality.

The exclusion of motor vehicles from the historic centres of cities, to mention but a single instance, involves the same questions as their exclusion for other reasons.

Greece, France and Italy offer a significant body of experience in the preservation of historic districts. Such war-devastated centres as Warsaw and Prague have usefully raised other questions. These the author of this volume has approached both from personal experience as a practising architect and city-planner (he is a member of the *Freie Planungsgruppe*, a consortium of young German architects and planners) and in a more academic spirit as a professor at the Berlin Technical University's school of landscape design. His ten years of professional practice in Greece is particularly well reflected in the present volume. Here Papageorgiou developed his fundamental view of the need to structure the design of historic areas in terms of its relation to the surrounding city, to its landscape and to the contemporary needs of the tourist traffic, which is at once the principal function to be served by the area and the origin of the contemporary preservation problem.

Everyone interested in historic areas has sat through endless meetings in which subjects were selected at random or to reflect personal or local interests, in which terms were never defined, in which choices never had to be faced, and in which conclusions were never reached because there was never agreement on what was important. Such vagueness can hardly be indulged when the cultural patrimony is melting away. The only remedy is the more rigorous approach offered in this volume.

But the subject rewards still broader exploration, as was made evident at the recent meeting of ICOMOS, the UNESCO agency concerned with historic preservation. Were time and space available, one would like to pursue further contributions to historic area preservation from the parks movement in English-speaking countries, from traffic management and from other fields of experience that appear of world-wide application. In nations like Japan an important recent awakening to historic area conservation is yielding both limited accomplishments and significant broader influences on contemporary architectural design. Perhaps it is in the English-speaking world, however, that the most dramatic change is to be seen. This falls outside the scope of Papageorgiou's treatment of the subject of historic centres of cities, but I believe the readers of this important book will be served by some description of this development, as well as by an introduction to the subject of the book and some indication of its importance to British, American and Canadian readers.

The recognition of historic centres of cities has made astonishing progress in the English-speaking world in very recent years. A decade ago this concept was unknown to the public mind, except in the case of museum-cities like Colonial Williamsburg. Little was being done to identify the cultural values they embodied. In Bath or Savannah, Edinburgh or San Juan, these were taken to be self-evident. Almost nothing was being done to preserve or enhance them. What preservation efforts were being made were narrowly historical in their objectives. Their aim was to save historic and architecturally significant buildings. These values were considered absolute by the small numbers of specialists who strove valiantly for their preservation. Only later was it seen that they were of vital importance to cities locked in a desperate struggle for survival. Only later did it become known that the natural setting of many cities, their river-fronts, terrain (like Mount Royal, which gave its name to the Quebec metropolis) and other features were equally significant to the urban environment. Only much later was it seen that both the protection of historic centres from traffic and other massive technological changes, and the further development by appropriate means of such centres to enable them to maintain their role as attractive living areas, was essential to their survival. These lessons have been learned from experience rather than from theory. It is perhaps characteristic of British and American attitudes that some general theory of historic centre preservation, as is offered in the present volume, should come after some significant work in the field has been done rather than before. We want to explain what we did, not to plan ahead. Like the famous bird that flew backwards, we would rather see where we have been than where we are going. This is another way of saying that the author, who is qualified by experience as well as by his academic background and research, addresses himself to English readers who are interested in this topic, at the right time.

Great Britain turned from preserving individual monuments to entire districts in the early 1960s and under that remarkable influence, the Civic Trust. Not historicism but environmental quality became the objective. As it worked out, the Civic Amenities Act 1967 required local planning authorities to designate 'Conservation Areas' and to enhance their character. Before, as a means of discovering how conservation policies might be implemented, the Minister of Housing and Local Government and the City and County Councils concerned jointly commissioned studies in 1966 of four historic towns: Bath, Chester, Chichester and

York. The purpose of these studies was to produce solutions for specific local problems and to develop policies for general application to all historic towns. Since all four towns had previously been the object of much preservation effort, mainly directed at individual buildings, they offered a specific opportunity to illustrate the application of newer amenity values.

These studies have exceptional value as illustrations of planning methodology. They have been summarized in the following pages in order to indicate the British approach to historic district preservation and to suggest how much these expert, preliminary planning exercises have to contribute to the world-wide preservation movement. Apart from their specific relevance to the Civic Amenities programme, the British awareness of urban vitality, continuity and identity is striking; and the broader-than-architectural interest in both the preservation of scenic, folkloric, commercial and other values and in the development of graphic design, street furniture and other unifying and strengthening activities is full of meaning to other places. These historic districts have contemporary value. They are not museums. They are meant to be lived in, to be used and to be enjoyed.

The historic towns studies were conducted by individual consultants and were designed to offer an experimental variety of planning approaches and solutions within a general structure of comparability.

The study of Bath (Colin Buchanan and Partners) focused on the early architectural heritage of the town, prior to its great eighteenth-century period. This 'subtle blend of purely architectural and human values' challenged the development of new and more comprehensive measures of preservation than had been applied to saving structures alone. New survey techniques were also developed. From the survey it was determined that many of the most important buildings were in poor condition, much of the space they contained was unused, and most of the buildings were poorly suited to contemporary use because of poor internal arrangements and bad external environment. This suggested a solution in which façades were saved but considerable reorganization and renovation of interior spaces was to be undertaken. Substantial upgrading of the landscape and townscape elements of the area were also proposed. New traffic regulations to protect the area and finding new uses for buildings (such as for student housing in the new University of Bath) would meet the major problems in the area. However, the costs of such measures required the most careful organization of public and private efforts, a challenge to the municipal corporation itself.

The Chester study (Donald W. Insall and Associates) examined the town in the larger regional context which holds the major promise for strengthening the older central core. The town centre was defined in all its two thousand years of architectural styles and social patterns, surrounded by the river, the walls and the canal. Proposals for further development of these features offer recreational and tourist values as well as a strengthened image of the historic centre with its Roman camp plan. A walk on the walls here, as in some other towns, would yield a revelation of identity, and could be one of the city's greatest attractions. In greater detail, more than 400 buildings were evaluated. Many were found seriously decayed. Apart from the recommendations on traffic and environmental improvements, and the designation of the area within the walls as a Conservation Area under the provisions of the Civic Amenities Act 1967, the consultant further recommended measures for the preservation and adaptive use of the old buildings. Pilot schemes could be carried on through the City Corporation, as owner or in consortium with private developers, redevelopment by housing societies, or local redevelopment by national organizations. The orchestration of many sources of assistance and their relation to a variety of problems was recognized. A new national agency, the Historic Towns Corporation, utilizing the existing powers of the Land Commission and other grants, loans and incentives to encourage conservation, was proposed.

Chichester (G.S. Burrows, consultant) was approached as a regional planning problem created in large part by recent accelerated growth as a recreational, yachting, cultural and residential centre in West Sussex. The visual identity of this cathedral town and its unique amalgam of historical styles has produced an extremely pleasant environment impossible to duplicate under modern development conditions. The study commenced by establishing the genuine threat to the character of the town and the steady loss of listed historic buildings. The inventory also recognized a number of successful conversions of important buildings and other preservation, and the outstanding work of the town council in regulating outdoor advertising. The analysis also embraced Chichester's 1966 Development Plan and some subsequent proposals of importance to conservation, such as the introduction of pedestrian precincts, a mini-bus service and other traffic control measures. A depth analysis of an illustrative eighteen-acre study area included a townscape survey. The consultant's recommendations included a proposal to form Conservation Trust Associations to allow owners and occupiers to be more closely involved with the process of conservation and thus overcome the fragmentation of ownership.

That most medieval of all English cities, York (Lord Esher, consultant) was found to be experiencing a steady loss of listed historic buildings despite a very high level of municipal expenditure for preservation. After assessing the general character of the walled historic sector, the York study examined the human activities of the core, first by means of a sketch by a resident of a typical day in the life of the walled city, and second by an analysis of the core as a commercial and employment centre. This led to a diagnosis that central York, like other city centres, had slowly degenerated into a place to work and the exodus of residents had accelerated. Environmental improvements to make the area attractive for residential uses thus became a primary objective if the continuity of the city was to be assured through strengthening its economic and social vitality. This programme was illustrated in detailed studies of four areas. In this it became clear that residential uses were in conflict with commercial demands for roads and parking, and further conflicts were noted with educational uses and tourism. The consultant's recommendations reflected his new criteria and the priorities needed for successful preservation. Traffic diversion and controlled access were regarded as fundamental to produce the lower traffic levels that made pedestrian traffic agreeable. In this programme some low-priority uses that generated irrelevant traffic were to be moved to outlying areas. The residential population of the walled city was to be doubled – to 6,000. Some university functions were proposed for location in the centre, and tourism was to be diversified and enriched. A thoughtful and extended analysis of costs and benefits of conservation was offered, giving particular emphasis to the long-term staging appropriate to such activity.

Brief summaries of lengthy and detailed technical reports can do no more than suggest the imaginative resourcefulness of the original documents and their widespread application to many other cities.

The most threatening factor in historic towns and areas is traffic. This is obliging consideration of more comprehensive forces outside historic areas that are having an impact on its conservation. Here the main planning strategy is dictated by Colin Buchanan's report, *Traffic in Towns*, which advanced the theory of environmental capacity. When an area's capacity to absorb traffic is exceeded, pressures appear to demolish buildings to create parking, to widen streets and otherwise to accommodate increased traffic.

Simply the noise, vibration, congestion and hazards of increasing traffic in narrow streets destroy amenities in historic districts and reduce their appeal and viability to both residents and tourists. Here the conflict between local business interests demanding access and the low traffic requirements of preservationists conflict. While each area demands special study and its own plan, a strategy is emerging to limit traffic on existing streets of historic areas while providing reasonable access from the rear of business properties where special access streets and parking are planned. The town of Hatfield has developed such a plan. Closing some streets altogether has also been attempted. Norwich has joined more than one hundred European towns that have specially designated 'foot streets' (a rough translation of the German *Fussgangstrassen*). More British towns are following.

In addition to the specific local findings and recommendations of the consultants, important new ideas about historic district preservation feasibility and techniques are offered in their reports. Many of these are the work of economic consultants, traffic specialists and others mobilized to contribute to the preservation effort. This rich creative contribution leaves no doubt that large untapped creative and intellectual resources are available to support the preservation movement. It is not enough to rely on conventional practices.

In the aggregate, the four studies accurately anticipated one of the major problems encountered by the Civic Amenities Act 1967 – the rather weak capability of local governments in Great Britain to undertake and execute a sophisticated new programme of this sort. Even when analyses and plans by expert consultants have been prepared, there remain difficult problems of carrying out such programmes and recommendations. Stronger measures of coordination are required. New categories of local government personnel must be established and trained. New political and administrative instruments must be forged. A new climate of public opinion must be created in which to sustain the necessary employment of public powers. A better appreciation of the values to be won from conservation and amenity planning must support fiscal changes and the commitment of local tax revenue to this new purpose. The most difficult problems of civic amenities are those of housing and urban transportation – and no city has solved those.

Experience with the civic amenities programme has also fed back the message that local effort will not be enough. In many cases, national policies in many fields, including housing and transportation, must be changed to reflect the higher priority given amenity and environmental quality.

The British philosophy of historic area preservation is not simply to save the buildings and characteristic spaces and features, but to use them in the creation of a new environment consistent with contemporary interests and desires. These are too numerous to detail here, but it would certainly be correct to say that the civic amenities district will be healthier, quieter, safer, more responsive to self-conscious cultural interests, more respectful of the past, more livable and in many other respects distinctively more 'modern' than its original in medieval or Georgian times. It does not intend to recreate the old but to create something that is contemporary in its values as well as its appearance, much of it based on conserving the surviving old buildings and utilizing them freshly. At bottom this is both a response to popular appreciation of the inherent attraction of historic districts as a place for living and for certain kinds of business, and a strategy for their preservation by developing them to standards competitive with, but hardly identical to what would be built today.

The Civic Amenities Act may be killing Britain's architecturally interesting towns with a mistaken kindness. What seems required, on the evidence of the four studies of historic towns, is spending less on preserving individual buildings in order to spend more on traffic control and other measures that would have greater leverage and accomplish the larger objectives of area conservation and amenity. But this is seldom what preservationists want.

In the United States the legislative landmark was the 1966 Historic Preservation Act. This separates the earlier Federal landmarks and registration programmes administered by the National Parks Service and a miscellany of state, local and private efforts, from the broader preservation goals and grants-in-aid programmes then inaugurated. Despite this great strengthening, preservation efforts in the United States remain scattered among governmental and private agencies, operating on the federal, state and local levels. Today there is at least a single focus, the Department of the Interior's Office of Archeology and Historic Preservation, which administers the National Landmarks Commission, the National Register of Historic Places and the Historic American Buildings Survey. It also embraces an Advisory Council on Historic Preservation which is empowered 'to comment upon all undertakings licensed, assisted, or carried out by the Federal Government that have an effect upon properties in the National Register'.

The rationale for the 1966 legislation was enunciated in the volume *With Heritage So Rich* by the Special Committee on Historic Preservation. The Committee concluded that preservation efforts must recognize the importance of 'architecture, design and aesthetics as well as historic and cultural values', and that preservation must concern itself with historic and architecturally valued districts having 'special meaning for the community'. As is deemed appropriate for a federal government with limited powers, little in the way of direct national programmes is undertaken in the United States, but increasing emphasis is given to grants for such purposes as the undertaking of state-wide historic site surveys, preservation plans and to individual projects for 'acquisition, protection, rehabilitation, restoration and reconstruction' which are undertaken as part of approved state-wide preservation planning.

As of February 1970 there were sixty-three historic districts on the National Register, forty-one of which had also been designated National Historic Landmarks. Formal recognition by the National Park Service of outstanding national importance is required for designation as a National Historic Landmark. In addition to the historic districts listed on the National Register, there are approximately one hundred districts on state and local registers.

The use of zoning to preserve historic districts began in 1924 in Charleston, S.C. and in New Orleans, La., which were the only cities in the United States to have such zoning until 1946, when Alexandria, Va. enacted its historic district zoning ordinances, and approximately 190 additional communities are contemplating such legislation.

Historic district zoning legislation generally sets up an Historic Districts Commission, whose approval is necessary for the alteration of exterior architectural features visible to the public and for the demolition of structures within the district. Thus, administration of the district rests in local hands. A second type of historic district legislation derives from the development of a centralized programme operating at the state level. The 1966 National Historic Preservation Act encourages this approach. Hence, responsibility for restoration and maintenance falls on private investors.

In spite of the legal and economic entanglements issuing from the public versus private spheres of power, it becomes clear that additional controls will be needed to protect the environmental quality of the increasing number of historic districts in the United States. A starting point would be the development of environmental standards on such matters as street landscaping, parking and traffic problems, maintenance of vistas, open areas and scenic amenities, and gardens and plantings. Protection from destructive intrusions and removal of incompatible uses must be part of future preservation programmes. At the same time, activities that would revitalize historic city centres must be identified and encouraged.

Agencies other than the National Parks Service concerned with historic preservation
In addition to the National Parks Service there are various other agencies concerned with preservation. A recent count of federal agencies having programmes affecting preservation listed fourteen departments (including the Department of the Interior) plus the Executive office of the President, in the Executive Branch of the Federal government alone.

The Department of Housing and Urban Development operates three programmes that are directly related to preservation activities. The smallest operating programme provides direct grants for historic preservation of up to fifty percent for acquisition, restoration and development. Under the Urban Renewal Administration there are two historic preservation programmes. Since the 1966 legislation, historic preservation activities have been included in the costs of urban renewal projects. Federal funding is provided on a three-quarter basis for cities with populations under 50,000, and two-thirds for cities over 50,000. Up to $ 90,000 can be provided by the Federal government for restoration of the exterior and interior of an individual building, if it is open to the public on a frequent and regular basis. The largest and most important current project is the Boston Waterfront area, where forty-four buildings on N. and S. Market Street are being preserved through Federal funds, totalling approximately two million dollars.

A second activity of the Urban Renewal Administration is the Demonstration Programme, which was organized for the purpose of eliminating slums and blight in urban areas. One major study has been made of historic districts, focused on the College Hill area of Providence, Rhode Island. Part of the purpose of the College Hill report was to correct the piecemeal approach to historic preservation in this country by developing a comprehensive approach to renewal of an historic area that could serve as a guide to other cities.

The Department of Transportation is another major federal agency whose activities directly affect historic preservation. DOT in the past has been notorious for its destructive indifference to other than a narrow range of engineering and economic criteria. Recently, however, it has made a diligent effort to use federal powers to influence preservation on the state and local levels. Under section 4F of the 1966 Department of Transportation Act, a new national policy was announced requiring that a special effort be made to preserve historic sites. The Secretary of Transportation cannot approve any project that would affect properties of local, state or federal interest if a feasible and prudent alternative exists. All possible planning efforts must be made to minimize harm to historic properties. The most significant action under this new legislation was the cancellation of the riverfront expressway that would have destroyed part of the Vieux Carré historic district in New Orleans.

Without doubt serious questions about social benefits face the preservation movement in American cities. Saddled with an archaic system of public finance, cities have neither the resources necessary to escape hard choices about their priorities, nor the organizational ability to resolve these conflicting choices. Federal programmes are badly fragmented. The movement to preserve historic districts stands in the shadow of earlier efforts to save individual buildings. These have almost universally been responsive to middle-class values and manned by middle-class people. The larger issues of district preservation are as complex as housing, urban renewal or urban expressway building. They threaten relocation of low-income families. They appear indifferent to black residents and the problems of employment and housing they face when dislocated. When city councils are faced with historic preservation they must ask what the realistic objections may be. It is not important that such objections may be mistaken, prejudiced or even self-seeking. The point is that they are reflected in votes and must be duly weighed in the political process. Considerable leadership must be displayed if cities are to save their historic centres, their individual identities and – to put it plainly – if they are to save themselves in the competition with suburbs, with other cities, and with their own change.

Already it is clear that the importance as well as the difficulties of historic district preservation have been underestimated. Much more careful investigations of historic areas, more skilful economic analysis, more sophisticated and imaginative planning, better and more sensitive architectural design – and more – are needed. And it all costs more money. But equally there can be no doubt that in every case which has been examined thus far, the benefits outweigh the costs. Perhaps some description of these benefits will bring this introduction to a suitable conclusion.

New forms of governmental organization must be developed, suited to the special tasks of historic district preservation and use – as new as port authorities, turnpike commissions, government-owned corporations, 'Comsat-type' mixed enterprises, or what is called 'multiple-purpose-planning-joint-development'. New forms of private enterprise, equal to the limited dividend corporation, the private foundation, the research and development corporation or the condominium, are also needed. Such institutional innovation may desirably extend to the realm of citizen participation, voluntary associations, block or precinct action groups.

Historic district planning in the United States has yet to achieve its urban focus. Relatively few districts are truly historic centres of their towns, but those are of the greatest significance. Annapolis, Charleston and Newport are all excellent examples of districts of major cultural and architectural importance. Nor are there many like the Vieux Carré, College Hill and Brooklyn Heights urban districts where an architectural style and period are notable. It is still historic rather than architectural values which predominate. Mining towns, mill towns, whaling ports, communities of ethnic or religious character – these make up most of the American historic districts. They must be museum towns, carefully and even reverently preserved. They cannot be vital historic centres, giving individuality and character to entire cities and metropolitan areas, and generating new urban strength in the effort to sustain urban life. What might have been the true historic centres of American towns and cities have been lost in the tides of economic and social change. To review the wasted treasures in the Stokes collection is a sombre experience. Much of the weakness of American cities, their failure to maintain their unity and centrality, their lost local patriotism and loyalty, can be traced to the disappearance of this identity, much of it in very recent times and because of frantic and misguided efforts at 'urban renewal' to lure suburban families back to town and the building of urban expressways thought likely to strengthen the old central business district.

In English-speaking countries there appears to be agreement on what an historic district is, on how it might be viewed, and on what its values are. There is also agreement on how such districts should be analysed and what their principal problems are. It also seems clear that such districts require modernization if they are to continue to be useful, and that preservation therefore demands some measure of development. The principal issues that must be faced are the modernization of housing, improved traffic circulation and parking, and the incorporation into the contemporary life of such districts of services and commercial facilities both for their residents and for tourists. What has created division – and a useful diversity of effort – is how such new development can best be incorporated into historic districts, given the conservation of their cultural values. We are close to the time when a book can be written on this subject, but that is not the aim of the present volume.

INTRODUCTION

The principle of protecting and preserving architectural monuments as emblems of our cultural heritage was established during the first half of the nineteenth century and gradually gained public support. But it was not until a good hundred years later that the idea of preserving and regenerating complete urban centres, which possessed historic or aesthetic value, began to assume tangible form. Subsequently, in the course of our own century, the destruction wrought by two world wars acted as a catalyst and produced a new and more informed approach to this problem.

During the reconstruction of the historic urban centres damaged in the war, new methods and techniques were evolved for the preservation of buildings. It also became apparent at that time that the increasing urbanization taking place in Europe and the radical change that had been effected in the scale and structure of the modern urban cluster posed a serious threat to the very existence of historic centres. This then led to the development of a special branch of research into town-planning that is concerned primarily with the protection, rehabilitation and regeneration of historic urban centres.

This venture has often been subjected to harsh criticism, consisting largely of accusations to the effect that it is a completely illusory undertaking based on anachronistic, romantic and eclectic principles. The main argument put forward by the anti-preservationists is the fact that, whenever man has created original works of architecture, he has done so by turning away from traditional concepts.

This is, of course, a perfectly valid point, but it does not invalidate the case for the preservation of historic towns for, ever since the onset of the industrial revolution – which set up its own aesthetics and its own style of architecture – there has been a binding obligation on us to adopt a more positive attitude to our cultural heritage.

It is significant that the movement for the preservation of individual monuments was launched between 1820 and 1850, when the new industrial society was first beginning to assert itself, while the movement for the preservation of whole urban centres, which was the next major step taken in this direction, was started in our own day and coincided with the emergence of prefabricated and mobile architecture.

It gradually became apparent that the new developments taking place within the spheres of architecture and town planning were so radical that exceptional measures would have to be taken as a matter of urgency if the historic urban centres were to be preserved. Otherwise, within a few decades, they would either be throttled by urban congestion or ruined by land speculators, or else their character would be completely destroyed by the intrusion of unplanned new buildings.

Today we are acquiring a new 'awareness of historic values' due largely to the justifiable misgivings which we feel about allowing the whole of our architectural heritage to be demolished now that we are entering the new era of the flexible and mobile urban cluster.

In actual fact, every creative epoch has produced its own individual conception of architecture and of town planning. The present epoch is no exception. Our architects and town-planners are pursuing the same revolutionary path as their predecessors. But today their pioneering zeal is tempered by the realization that in any new projects allowance must be made for the preservation and regeneration of our cultural heritage.

Consequently, any undertakings designed to ensure the survival of our historic urban centres, far from undermining the achievements of modern architects, actually contribute to their activities in so far as they help to promote the integration of these traditional structures and to define their role in the urban cluster of the future.

This axiom is disconcerting, both to those who mourn for the familiar townscape of the pre-industrial era and to those narrow-minded and fanatical prophets of a new world in which there will be no historical continuity and no architectural heritage.

Meanwhile, however, the problems of urban regeneration have roused the interest of talented architects and informed laymen in many countries. Groups of architects have carried out comprehensive studies of important, historic centres such as Regensburg in Germany and the Marais quarter in Paris. Moreover, individual historic buildings are now being reconstructed, converted or regenerated in ever increasing numbers.

This present work was prompted by the author's preoccupation with 'Plaka', the 'old town' of his native city of Athens. His interest was first roused by the conversion of an old patrician house of the 'Kalifronas' in Plaka. This led him to undertake a detailed analysis of the architectural and planning problems facing this historic sector, as a result of which he was obliged to come to terms with certain basic questions regarding the validity and feasibility of regenerating historic centres and to consider the kind of techniques required for such an undertaking.

This book may therefore be regarded as a European contribution to the new discipline which has developed within the general sphere of town planning.

The author wishes to express his profound gratitude to his late father Nikolaos Papageorgiou, who first taught him to appreciate the visual quality of the environment, and to his teacher, the late Professor Demetrius Pikionis, from whom he learnt to observe the world with an 'inner eye'.

The author also welcomes this opportunity of thanking all those who have contributed to the writing of this book.

Professor Herta Hammerbacher of the Technical University in Berlin, upon whose recommendation the author was granted a two-year research fellowship in Berlin by the Alexander von Humboldt-Stiftung in Bad Godesberg, was responsible for initiating the work, which then received further impetus from the generous interest taken by two members of the Stiftung, Dr. Heinrich Pfeifer and Dr. Thomas Berberich. Professor Kurt Bittel, President of the German Archaeological Institute, Professor H. Hammerbacher, Professor F. A. Gunkel, Professor H. Reuther, Dr. F. Mielke and Dipl.-Ing. M. Korda kindly read the manuscript and made various useful suggestions.

Frau Margot Larisch of Bad Goedesberg afforded valuable help with proof-reading.

The publishing houses of Vincent et Fréal in Paris, Ernst Wasmuth in Tübingen and Praeger in New York, who have collaborated in publishing this work in three languages, have been extremely cooperative.

To all of these people the author wishes to express his sincere thanks.

West Berlin, March 1968 – June 1969

1 Historic urban centres

1 Historic urban centres

11 GENERAL APPROACH

111 *Historic Urban Centres within the Context of the Present Study*

During the past few decades a new branch of town-planning has emerged. This discipline, whose theoretical basis has been gradually built up over a period of more than a hundred years, is concerned with the analysis of all aspects of historic centres[1] as these appear to us today within the framework of the artificial cluster. Apart from providing definitions of concepts, the research in this field has also considered the aesthetic problems involved by preservation and has made practical suggestions for the integration of historic urban centres.

This present study constitutes an attempt to investigate the kind of problems raised by such research.

First we propose to define the concepts 'historic urban centres' and 'urban regeneration', 'sanitation' and 'rehabilitation'.

Then we shall try to clarify the following aesthetic processes:

a) The development of the concept of the 'historic urban centre' and the consistent attempts undertaken to ensure its survival;

b) The present significance of the historic urban centres within the framework of the modern urban cluster and its corresponding 'townscape';

c) The future significance of the historic urban centre within the future spatial cluster and its corresponding 'spatial setting'.

And finally we shall describe methods of procedure to be employed in architectural and town-planning projects involving historic sectors and towns.

---

1 The terms 'historic towns', 'historic centres', 'historic urban centres', 'historic settlements' and 'protected urban areas' have been treated as synonyms for the purposes of this study. They all denote 'living', i.e. inhabited, settlements of considerable aesthetic or historic importance. For a more precise definition of this concept, see sections 121 and 122.

1

The forum of Pompeii. This Roman town, which was buried under lava when Vesuvius erupted in A.D. 79, has been almost completely excavated. The excavations were started in 1748 and continued virtually right down to our own day. Pompeii is in an almost perfect state of preservation and there is little to remind the visitor that this is an archaeological site. The residential quarters look as if the population had only just moved out. None the less, Pompeii cannot be regarded as a historic urban centre and must be classified as an archaeological site since it lacks one of the principal attributes possessed by every historic town, namely urban activity arising out of social life.

The investigation, description and classification of historic sources, remains and documents is undertaken by a whole group of related disciplines. Archaeology, 'museology', political history, the history of art, the history of science, sociology and the history of town-planning are all concerned with the analysis and description of the phenomenon of human development in the fullest anthropological sense of the word. Moreover, these studies are not prompted simply by scientific curiosity. They are also intended to provide a model of man's future development on earth.

However, none of these disciplines regard historic objects or 'remains' as potential 'carriers of life', i.e. as a framework for present or future human activity. On the contrary, they treat all such objects from the past simply as the component elements of historical records.

Our new town-planning discipline is quite different in this respect. It is not simply intent on documenting historic towns. On the contrary, it regards them from the dynamic viewpoint of their present and future development and treats them as carriers of present and future life.

12 THE CONCEPT OF THE 'HISTORIC URBAN CENTRE'

121 *Meaning and Scope of the Concept*

For a town or urban sector[2] to be regarded as a 'historic settlement', it must possess:

a) *An original and characteristic urban structure (originality of the composition);*

b) *Significant architectural qualities (architectural monuments and interesting buildings) whose structure points to a marked degree of continuity in the urban development of the settlement (aesthetic and historic value of the composition);*

c) *A continuing social life, i.e. some form of civic activity, which presupposes the existence of an active population ('living' condition of the composition).*

From this it follows that:

a) Contrary to common belief, historic settlements do not have to be very old. The attribute 'historic' refers to the *whole* historical development of the settlement and not simply to its origin in time. Consequently, interesting urban formations of quite recent times may also be classified as 'historic settlements';

b) Since the designation of a settlement as a historic centre depends on the existence there of an active social life, excavated sites and archaeological remains and even 'dead towns' whose buildings have been extremely well preserved (e.g. Pompeii) cannot be regarded as historic towns. Abandoned settlements form part of the subject matter of archaeology and the history of architecture, but they do not enter into the sphere of town-planning.

c) Since many inhabited settlements today do not possess an 'original urban structure' and lack 'aesthetic value', it naturally follows that not all urban settlements can be classified as historic.[3]

---

2 The different categories of historic urban centres are presented in section 132.

3 Consequently, we find ourselves today faced with the task of compiling or completing national and international inventories of all those settlements which fulfil the conditions outlined above. The assessment of these settlements would need to take account of their aesthetic and historical value, their architectural quality and their present function within the urban cluster.

In compiling national inventories of historic centres, we shall have to consider further criteria which are quite as subtle and complex as those already dealt with:

a) The unique character of every urban composition. Not even in the case of small garrison towns, which were all planned as military strongholds, do we find completely identical urban formations. The particular character of every urban formation is, of course, determined in the first instance by its geographical location, which naturally makes for uniqueness. On the other hand, we often find settlements which, although not identical, were built by the same society, at the same time and in the same architectural style. Moreover, many of these have subsequently passed through a similar historical development. In certain areas (e.g. on the islands of the Aegean), there are large numbers of small towns of this kind. But the presence in a particular area of different settlements of basically the same type should not lead us to accord special treatment to some at the expense of the others.

On the other hand, we have to ask ourselves whether all such 'similar' centres should be preserved or whether we ought to content ourselves with a representative selection and concentrate our energies on these.

b) This raises the difficult question of priorities and *in this respect, it must be stressed that the significance of a historic settlement transcends the antiquarian value of its architectural monuments and must be sought in the totality of the urban cluster (see 4111). Basically, all historic centres are equally significant and, in principle at least, there can be no question of establishing priorities within the sphere of preservation.*

Nonetheless, if we consider this problem in practical terms, then in view of the large number of urban centres in need of rehabilitation and the great amount of money needed to carry out a full scale programme, we are obliged to set up a scale of priorities.

The criteria that would govern any such selective classification are, firstly, *the uniqueness of the urban composition* and, secondly, *the assessment of the architectural quality and the geographical locality of the historic centre in the general urban cluster.*

Having made this practical concession, however, it must be said that if we are to adhere strictly to the concept of the historic centre outlined above (see 121–122), then the overwhelming majority of existing urban centres would have to be classified as 'historic' and consequently worthy of rehabilitation.

From this it is clear that the rehabilitation of historic urban formations poses a social and planning problem of considerable magnitude. In the first phase of any rehabilitation programme, we would need to concentrate on the most interesting and valuable historic settlements. The final aim, however, would be the regeneration of all historic settlements within the artificial cluster.

2
A historic urban centre that is full of life and rich in historic and architectural monuments: the Bazaar of Constantinople with the Suleiman and the Zadeh mosques in the background.

3
An example of a historic urban centre that was badly damaged in the Second World War: the inner town of Hanover.

4
An example of a historic urban cluster that has been perfectly preserved: the Binnenhof in The Hague. In the centre the towering structure of the Rittersaal, which is now used as a throne-room in the ceremonies which accompany the opening of the Dutch parliament.

131   *General Remarks. The Survival of Historic Towns*

Until recently the whole of Europe – from Spain to Sweden and from Ireland to Greece – was covered by a complete network of historic centres which existed side by side with twentieth-century towns and suburbs. The majority of these settlements had remained virtually intact, having survived for centuries on end in a quite remarkable way. There were, of course, exceptions to this general rule, including St. Mark's Square in Venice, where the campanile was destroyed by an earthquake in 1907, Heidelberg Castle, which was burnt down by French troops in the eighteenth century, and the Tuileries, which were gutted by fire as a result of the fighting during the Commune in Paris.

But then, in two world wars, the whole of Europe was virtually laid waste, with the result that its historic townscapes underwent a radical and permanent change.

Meanwhile, however, the growing awareness of the cultural importance of individual architectural monuments and of groups of buildings within historic centres prompted a powerful response, not only from the specialists in this field but also from the broad mass of the people. As a result those countries whose towns and cities had suffered in the fighting set about the regeneration of their historic urban centres with great energy. However, although some of these reconstruction projects were highly successful, others were not.

But despite wars, desecration and the changes wrought by the unplanned expansion of big cities (resulting in the demolition of buildings in historic urban centres to make way for new roads to carry motorized transport), a large number of historic centres have survived *in western Europe due to quite specific historical*, *geographical*, *political and economic circumstances*, which may be subsumed as follows:

a) The political fragmentation of large areas of western Europe, which set in from the end of the Middle Ages, promoted the growth of large numbers of centres in the form of feudal capitals, free cities (Hanseatic and Italian cities) and royal capitals (the petty principalities of Germany). Thus an extremely dense network of important towns was created at a very early stage. Moreover, these towns were all remarkable for their architectural quality and their ancient cultural traditions, which they have retained as historic cities right down to the present day. This is one of the reasons why there are so many – literally thousands – of historic settlements in western Europe.

b) As far as the social attitude of the inhabitants of these historic centres is concerned, we must bear in mind that a large proportion of them feel extremely close ties with their traditional townscape and tend to identify with the spatial setting of their historic towns. Consequently, they take a keen interest in all questions of sanitation and regeneration. Moreover, a large number of these inhabitants – who are relatively well educated – have a ready appreciation of the historical and aesthetic character of their ancient townscape and it is to be assumed that this also binds them to it.

c) During the past two hundred years, town-planning activities in these historic centres have been restricted largely to peripheral growth. There has been relatively little reconstruction in the town centres, due partly to the inhibiting effect of the monumental architecture and partly to the fact that these were already high density areas with buildings of between four and seven storeys.

d) Finally there is the fortunate circumstance that there are virtually no important archaeological sites within the precincts of the living historic towns of western Europe. Consequently, there is little or no likelihood that whole historic suburbs will be demolished in the interests of archaeological excavations. Unfortunately, the present situation in some countries is quite different. Because of the negative conditions obtaining in Greece, for example, only a limited number of historic settlements has been preserved and these are constantly exposed to a variety of dangers, which have arisen in the following ways:

a) The territory occupied by the modern state of Greece produced virtually no important administrative or cultural centres for a period of fifteen hundred years (Byzantine and Turkish eras). The only cities to flourish during that period were Salonica and – for a relatively short while between the thirteenth and fifteenth centuries – Mistra on the Peloponnese. Apart from these two places, Greece was one of the most destitute provinces in the Byzantine and, later, the Ottoman empire. Most of the great centres of the Byzantine empire, such as Constantinople, Dyrrachium, Ochris, Philippopolis, Nice, Antioch and Alexandria, were situated outside the borders of modern Greece. It is not surprising to find, therefore, that historic Greek centres include various small feudal towns and strongholds dating from the period of Franconian and Venetian rule (Corfu, Rhodes, Kandia, Chanéa), rural centres (which have furnished interesting examples of popular architecture) and, finally, nineteenth-century Neo-classical towns.

b) The relative ignorance of the broad mass of the Greek population concerning the aesthetic and cultural value of their 'living' architectural heritage, coupled with the complete lack of interest evinced by Greek researchers (art historians and archaeologists) in anything other than Byzantine or classical antiquities, has produced a climate in Greece that is distinctly unfavourable to the preservation of historic settlements. In fact, the living architectural tradition is regarded as a positive handicap and an obstacle to the 'reconstruction' of the few historic centres to have survived. (In the whole of modern Greece there are only about fifty such centres, as compared with two thousand in France.)

c) The basic structure of buildings in historic Greek towns also militates against preservation. These buildings are all rather low (one to three storeys), which has tempted Greek builders to demolish the old buildings in the urban centres so that, by rebuilding higher, they might put their sites to more profitable use. The radical redevelopment of the historic urban centre of Athens, which has been going on for the past thirty years and is now nearing completion, and the corresponding development in the Neo-classical centre of Patras are striking examples of this policy. Greek town-planning today involves not only peripheral growth, but the radical and speculative 'reconstruction' of historic urban centres, which are, of course, being destroyed in the process.

d) Finally, great difficulties are encountered in Greece due to the traditional practice of erecting new buildings on old sites. As a result living historic towns and urban sectors are often found to have been built on top of valuable archaeological remains. Choosing between these different cultural values poses an extremely difficult and critical problem. Unfortunately, the choice has all too frequently been made in favour of archaeological excavation, at the expense of the living urban centres (see 4114).

It should be pointed out, however, that of recent years there have been reassuring signs of a change of heart in this respect, which would indicate that Greece is at last beginning to realize the necessity for preserving and rehabilitating her historic urban centres. The fate of the old town of Athens on the northern slopes of the Acropolis has become the subject of lively public debate following a publicity campaign mounted primarily by Greek architects.

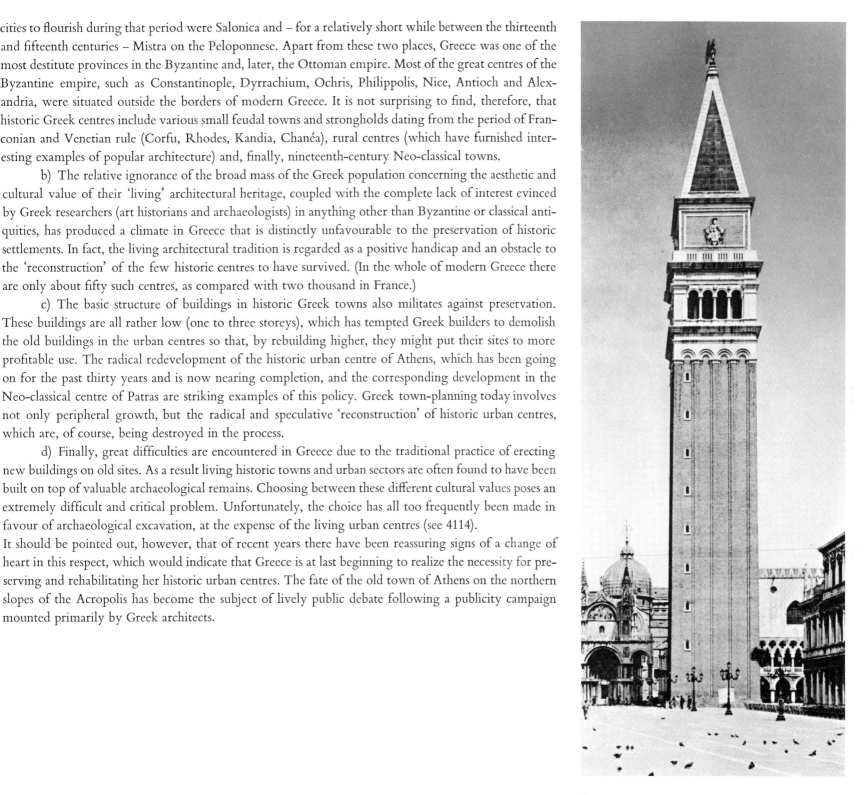

5

The campanile of St. Mark's in Venice, which was rebuilt in 1907 after the original tower had been destroyed in an earthquake. The symbolic importance of the campanile was one of the principal reasons for its arbitrary reconstruction.

6

Two-storeyed nineteenth-century burghers' houses on the main square of the town of Tripolis in the Peloponnese. Although morphologically interesting, these houses are very small and were built in a very simple way with low quality materials.

7

Façades of seven-storeyed houses on the quayside of the historic harbour of Honfleur in Normandy. These multistoreyed houses are morphologically interesting and were built with good quality materials on sites which were no larger than those used in Tripolis (fig. 6).

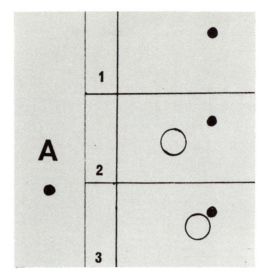

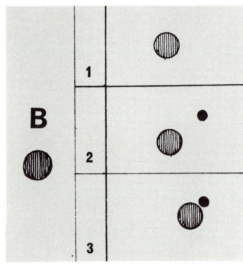

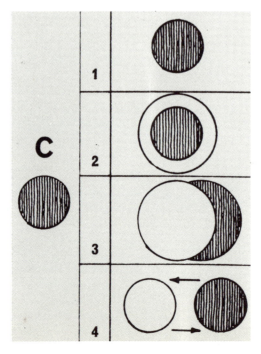

8

Diagram showing the different possible sites for a historic urban centre in geographical space:

a) Possible sites for a single group of monuments resembling a historic settlement (monasteries, castles, etc.);

b) Possible sites for a small rural historic centre;

c) Possible sites for a homogeneous historic town.

In classifying the historic centres of Europe in accordance with historical, geographical and morphological criteria, it would be possible to suggest a large number of categories whose analysis and comparison would then furnish ample scope for an independent study. But an analytical classification of this kind, which would be grounded primarily in architectural theory and in general history, would go far beyond the limits prescribed for this book, which is specifically concerned with the rehabilitation of historic urban centres and their integration into present and future 'spatial clusters' (see 222). Consequently, the classification presented here has been evolved strictly in accordance with town-planning criteria that are applicable to the conditions obtaining throughout the whole of Europe.

The historic centres of Europe may be divided into four principal groups on the basis of their importance as centres, their structure and their topographical position within the 'artificial cluster':

a) *Independent and monumental groups of buildings which resemble settlements;*

b) *Small rural historic centres;*

c) *Historic towns;*

d) *Historic sectors of big cities.*

a) The first group (see fig. 8 a) embraces historic monasteries and castles and a few historic farms (but only if they are inhabited). Many such groups reveal a wider and more interesting range of architectural qualities than are to be found in the small rural historic centres. However, we must regard this first group of historic centres as a marginal case since they have never been imbued with urban or social life in the strict sense. On the contrary, they have been centres of spiritual or patriarchal life.

As for their topographical position, we find these monumental groups of buildings in isolation[4], on the outskirts of small rural centres which possess no historic interest in themselves, in the immediate vicinity of historic rural centres or within the precincts of major historic towns.

b) The second group (see fig. 8 b) embraces all inhabited rural centres of historic and artistic importance. Such centres usually consist of small ports or of mountain and valley settlements. In fact, they are villages and small towns of little or no administrative, cultural or economic significance, whose relatively isolated geographical position has helped to maintain their urban structure unchanged for centuries on end, thus ensuring the preservation of their original, uniform townscape. Examples of such settlements include the small mountain towns of central Italy (such as Assisi, Urbino and Siena) and the villages on the Pelion peninsula in Greece. We find these small historic settlements either in isolation or in conjunction with one of the monumental groups of buildings listed under the first group, which may be situated either in the middle or on the outskirts of the rural centre.

c) The third group (see fig. 8 c) embraces all medium-sized historic centres (towns with a population of between ten and two hundred thousand) built in the last eight hundred years, which reached the peak of their development at some time in the past and then maintained virtually the same urban structure and the same sized population right down to the present day. These centres – most of them district or country towns – remained largely untouched by both the demographic explosion and the concentrated wave of urbanization of the past hundred years, with the result that their urban structure was never seriously impaired or altered. In some cases (e.g. the medieval cities of Regensburg and Chartres, the 'ideal' cities of the Renaissance and the Baroque city of Salzburg), the architecture and the townscape of these centres is homogeneous. In other cases (e.g. Florence and Venice), they are the product of successive creative epochs. Centres of this kind, which have evolved over the centuries, contain a mixture of styles. Some

9
Diagram showing the different possible sites for a historic urban sector in geographical space:

d) Possible sites for a historic urban sector in the heart of a major city.

---

4 Examples of monasteries which have been built in complete isolation but which are virtually self-contained urban communities include: Montecassino in Italy (which was completely destroyed in the Second World War), Mont-Saint-Michel in France and the monasteries on Mt. Athos in Greece.

historic towns have remained completely unchanged. Siena is a case in point. Due to its special geographical position, no new suburbs have been built in Siena and no attempt has been made to expand this town during the past two hundred years. Other towns – such as Florence – have merely retained their historic centres, which have been surrounded by new suburban belts. In a further group of historic towns, due to the presence of natural barriers, expansion has been restricted to just one side of the town (e.g. in Heidelberg and Ragusa), while in the case of historic towns situated on islands or peninsulas, new suburbs have frequently been built on the neighbouring mainland (e. g. Venice, Malvoisie).

d) The fourth and last group (see fig. 9) embraces the historic sectors of the metropolises of Europe. Today these centres consist either of isolated sectors within the urban cluster of our modern cities (as in Athens, Naples, Zürich and Geneva), or else they constitute an important part of the central metropolitan area (as in Paris, Amsterdam, Rome and Munich). There are five possible locations for these historic sectors within the metropolitan area:

1. Centre of the town;

2. On an important waterway passing through the town;

3. In the immediate vicinity of an important monumental group of buildings or a famous archaeological site within the inner metropolitan area;

4. In the immediate vicinity of an important green belt area within the urban cluster;

5. In the middle of other, newer sectors of the city.

At this point, reference should be made to those historic centres in big cities which have lost their historic structure over the past hundred years due to 'reconstruction' and the erection of new and incongruous buildings. A striking example of this regrettable development is afforded by the historic centre of Milan. By now it is too late to preserve the unity of Milan's traditional townscape, and all that can be done is to try to ensure that individual architectural monuments and historic squares are protected against further desecration in the future.

2 The Significance
   of Historic
   Centres
   within the
   Spatial Cluster

2 The Significance of Historic Centres within the Spatial Cluster

21 DEFINITION OF THE CONCEPTS 'ARCHITECTURAL MONUMENT' AND 'HISTORIC CENTRE' AND AN APPRECIATION OF THEIR CULTURAL SIGNIFICANCE

211 *The Early Stages*

The concepts 'architectural monument' and 'historic centre' assumed tangible form in the course of the past hundred years when, under the twin pressures of industrialization and urbanization, a new scientific, historical approach to town-planning was evolved, in which special importance was attached to the preservation of our cultural heritage (see Introduction).

Since the beginning of history, a wide variety of buildings and natural forms have been known and referred to as 'cultural' or 'natural' monuments. But, although these monuments were undoubtedly admired as objects of beauty and were also imbued with a certain historical value, their real significance lay in their religious or secular symbolism, which was far more important than their aesthetic or historical aspects. Consequently, it was a long time before effective measures were taken to protect and preserve these monuments.

The absence of protective legislation gave a free rein to the early architects, who did not hesitate to demolish old buildings of considerable architectural and historical importance in order to use the sites for new projects. The three successive temples erected on the Acropolis in Athens (the Hecatompedon and the old and new Parthenon), which were built between the seventh and the fifth centuries B.C., are a striking example of the way in which the Ancient Greeks rejected their architectural heritage in favour of new and original buildings.

But despite their absolute insistence on renewal and their disregard of traditional forms, these early architects evinced a profound respect for the 'sacred sites'; moreover, they invariably retained the traditional urban locations (sacred precincts, fortresses, agoras etc), thus preserving their original and important functions within the urban composition[5].

---

5 Both the town-planners and the art conservationists ought to ask themselves to what extent the symbolic, historic and cultural significance of such 'locations' (which often transcends the significance of the actual historic buildings) is protected in present-day redevelopment schemes.

Even in antiquity we find cases of peripheral growth. The new town of Athens, which was founded by Emperor Hadrian in the second century A.D., was built on the eastern side of classical Athens. The arch, which is still standing today and which bears the emperor's name, marked the boundary between these two urban areas[6].

It might be thought that the creation of a 'new town' outside the existing town centre constituted an early attempt to protect an entire historic settlement. But this would appear not to have been the case. In our view, the reasons for this innovation were of a more practical nature. It was partly in order to strengthen their political influence in Athens and partly in order to set up a rival to the old town that the Romans decided to build their satellite.

The first conscious attempts to legislate for the protection of architectural monuments were made during the last two centuries of the 'antique world'. In the cosmopolitan Rome of the fourth and fifth centuries A.D., whose population of one million was constantly increasing owing to the great influx of foreigners and barbarians from all parts of the known world, anarchy and civil strife were the order of the day. During that period, various emperors including Valens and Valentinian, Theodosius and his sons Honorius and Arcadius issued decrees designed to protect architectural monuments[7].

It is a remarkable fact that these measures, the first ever taken for the protection of architectural monuments, should have been evolved by such a powerful centralized state and that they were prompted by the threat of civil disorder in an overcrowded city.

What the Romans realized fifteen hundred years ago was that the real danger for monuments lies in neglect, in architectural deformation and in desecration brought about by arbitrary changes of function, a danger incidentally which is far more serious than that created by the actual destruction of buildings.

212   *Developments in Europe*

After neglecting their architectural monuments for centuries, the Italians were the first to take really positive steps to protect their heritage. This happened during the Renaissance.

The principle of the conservation of historic monuments was first established by Pope Pius II in a decree dated 18 April 1462 and was reinforced by subsequent papal decrees issued in 1516, 1655, 1701 and 1802. Meanwhile, the independent Italian cities also passed legislation to this effect.

Both the legal and the technical aspects of conservation were evolved at an early date in Italy, which possessed the most extensive Greek and Roman remains in western Europe, most of which were well preserved and had been integrated into the townscape of living historic towns. *This early realization of the need to preserve works of art was undoubtedly prompted both by humanist ideals and by the admiration felt during the Renaissance for the works of antiquity, which provided the inspiration for Italian architecture at that time.*

The first law for the preservation of historic works of art passed by the Italian government following the unification of the country ('Legge sulla conservazione dei monumenti e degli oggetti di antichità e d'arte' of 22 July 1902) was conceived in the conservationist tradition. It was an inadequate law in that the Italians, like the Greeks, were interested primarily in the archaeological remains of antiquity. Consequently, they

---

6 The 'Hadrian Arch' also bore two inscriptions, one on its west front facing the old town, the other on its east front facing the new 'satellite' town. These inscriptions defined the relations between the old and the new town. On the west side: ΑΙΔ'ΕΙΣ ΑΘΗΝΑΙ, ΘΗΣΕΩΣ Η ΠΡΙΝ ΠΟΛΙΣ, and on the east side: ΑΙΔ'ΕΙΣ ΑΘΗΝΑΙ ΑΔΡΙΑΝΟΥ ΟΥΧΙ ΘΗΣΕΩΣ ΠΟΛΙΣ.

7 One such decree, which was promulgated by Emperor Majorian in A.D. 458, contained the following injunction: 'The abuse, whereby men are at liberty to desecrate the image of the venerable city (Rome), must be done away with ... and so we command that any buildings erected by the Ancients for public use or embellishment, be they temples or other monuments, must not be destroyed or touched by anyone.'

were more concerned with organizing excavations and ensuring that their archaeological finds were protected. In Italy, immovable historic objects are protected partly by expropriation laws, which impose limitations on the property rights of private individuals, and partly by giving the state the first option in respect of all sales. A further effective conservationist measure taken by the Italian legislators is that prohibiting the export of works of art. This policy is understandable in a country like Italy which possesses a large proportion of the archaeological remains of antiquity since these are coveted by the whole of western Europe. It is also pursued in Greece for similar reasons but not in any of the other countries of western Europe, which all allow their works of art to be bought and sold in an open market. Italy was also fortunate in that she possessed a large number of important historic towns which were protected for centuries both from the ravages of war and from the repercussions of urbanization. But the extensive damage sustained by their historic buildings during the last war forced the Italians to face up to the critical problem of the conservation and restoration of their architectural heritage. They soon showed themselves to be particularly adept in this sphere, although at that time the problem of protecting whole urban centres had not yet become acute.

During the past twenty years, however, urbanization has speeded up considerably in towns like Genoa, Bologna, Florence and Naples due to the introduction of industrial sites and modern road networks, which threaten to destroy the unity of the traditional townscapes of these historic centres. On the other hand, Italian public opinion has now been made aware of such dangers. The flooding of Florence in 1966, and the gradual subsidence of Venice have provoked widespread interest in the fate of the historic towns.

Germany first realized the need to preserve her cultural heritage following the Napoleonic Wars. The fearful battles fought on German soil at the beginning of the nineteenth century was one of the precipitating factors; the other was the awakening of Germany's national 'conscience'. Sulpiz Boisserée and Karl Friedrich Schinkel, the master of Berlin Neo-classicism, were the first to call for the systematic protection of artistic monuments and in this they were supported by Goethe. But these early reactions were not calculated to produce concrete results; they were simply alarm signals sounded by men sensitive to art who were opposed to the many arbitrary actions taken by the government in a century that was full of unpleasant surprises[8].

Because Germany was divided up into more than fifteen states prior to 1870, it was not possible to evolve a unified programme for the protection of her cultural heritage. But it is interesting to note that the governments of the different principalities took early measures for the protection of various categories of monuments by issuing special decrees. Baden was particularly active in this sphere. So, too, was Bavaria, where the enlightened King Ludwig I reigned from 1825 to 1848. *One very important German characteristic, which has contributed greatly to the maturation of certain conservationist ideas, is the very early interest taken within the Germanic territories in the protection of the urban and rural 'townscape', and the image of the street (Strassenbild). The 'Heimatschutz' (protection of the homeland) is a specifically German conception which is found in no other European country. Strangely enough, this protective idea – which is concerned with both the German cultural heritage and with the German countryside – was developed in the course of the nineteenth century, i.e. before effective protective legislation was introduced. The 'Heimatschutz' also embraces customs, national costumes and rural and urban townscapes. It indicates an early awareness on the part of the Germans – one that is unique in the whole of Europe – of the need for protection, not only in the artistic sphere but also in the sphere of regional planning.* It is in nineteenth-century Germany that we find the first germ of the concept of the protection and rehabilitation of historic urban centres, which is now beginning to bear fruit.

---

8 For example: in Goslar, in 1820, the 'interesting cathedral so rich in memories' was sold by the government for demolition for 1507 talers.

Today the most advanced legislation produced in this sphere is to be found in the province of Schleswig-Holstein (passed in 1961; see Bibliography: D).

Unlike Italy, which was required to restore mainly individual architectural monuments following the Second World War, Germany had to face up to the more difficult task of rebuilding entire historic centres, for nearly all of her historic towns had been badly damaged. As a result, Germany became a testing-ground *par excellence* for every conceivable kind of reconstruction technique (see *in extenso* 4123). These experiments, which are described in Section 41 of this study, were immensely valuable. Some were successful, others were not, but they all drew attention to the problems involved and prompted a general discussion of the principles of reconstruction throughout the whole of Europe. Unfortunately, the absence of standard legislation and a certain lack of initiative on the part of the administration made it impossible to launch a coordinated programme for the rehabilitation of all the historic settlements in Germany.

It was left to France to make good this administrative deficiency. *While Italy's major contribution in the sphere of preservation has been the development of new conservation techniques for works of art, and while Germany acquired an early appreciation of the importance of the 'townscape' and, of more recent years, considerable expertise in the reconstruction of whole urban areas, France has always taken the lead in the establishment of new and effective administrative methods.*

The French authorities, backed by the specialists and by public opinion, took an early interest in the protection of individual living architectural monuments. The great wave of destruction unleashed by the Revolution, in which the French churches and monasteries were particularly hard hit[9], provoked a powerful reaction. In 1826, Victor Hugo called for 'war on the destroyers' and in 1837, the 'Commission Nationale des Monuments Historiques' was founded. The French Romantics rediscovered the Middle Ages, thus creating a general interest in this period, which led to the protection and restoration of many medieval buildings. In fact, the 1830s and 1840s saw the onset of a positive mania for restoration, which offered excellent chances of survival for France's historic monuments but which also produced occasional follies due to the excessive zeal of the restorers. One of the most gifted and prominent architects working in this field was Viollet-le-Duc. But, despite his undoubted talent and his thorough knowledge of historical building techniques, even his work was uneven. In the radical restoration of the Cathedral of Vézelay, which he carried out from 1840 onwards, he showed himself to be a conscientious and cautious restorer. But the conservation and extension of Notre Dame in Paris, which he undertook subsequently, is a far less conscientious piece of work. Unfortunately, in his late period, when Napoleon III commanded him to reconstruct the castle of Pierrefonds, Viollet-le-Duc created a *tour-de-force* which demonstrated his powers as an inventive but essentially frivolous restorer. On 30 March 1887, a law was passed which established different categories of historic monuments based on both scientific and legal criteria. This legislation was subsequently extended by a second law of 31 December 1913. Meanwhile, by the beginning of the twentieth century, some two thousand French monuments had been classified and listed in a national inventory. Today, this inventory contains over thirty thousand monuments. In France, monuments in private hands can only be classified with the owner's consent and, in the event of this being refused, the only course left to the authorities is expropriation. But, although a large number of French monuments are privately owned, cooperation between their owners and the local and national authorities has been good. Consequently, these monuments enjoy the same measure of protection accorded to those in public ownership. French architects began to specialize as 'custodians of monuments' over a hundred years ago when the 'Inspectorate of Historic Monuments' was established. This specialist body is concerned not only with the material conservation of monuments, but also with their functional survival in living environments.

10
Statue of the nineteenth-century French architect Viollet-le-Duc posing as a medieval master mason admiring his finished work. This statue formed part of the new sculptural decorations created for Notre Dame in Paris when le-Duc restored the cathedral in the middle of the nineteenth century.

---

9 The decree of 1792, which provided for the 'destruction of all monuments that recalled the *féodalité*' and 'all objects that evoked memories of despotism', testifies to the blind fanaticism with which the leaders of the Revolution assessed the cultural value of monuments.

The new law of 4 August 1962 – which is better known as the 'Malraux law' – and the directive of 13 July 1963 afford further examples of France's preeminence in the legal and administrative aspects of preservation. The directive provided for the establishment of 'protected urban sectors' within historic towns and laid down the methods of regeneration and rehabilitation (see *in extenso:* Techniques, Chaps. 42 and 43) to be pursued in respect of whole historic urban centres. There are two thousand historic towns in France and so far only forty protected urban sectors have been established. Special building plans and regulations have been evolved for these sectors in order to preserve their traditional character, and important sanitation works are already being carried out. In the sector of the 'Marais' in Paris, and of 'Saint-Jean' in Lyon, in Sarlat and in various other places, experimental techniques are being employed. The results of these experiments, which will give us some indication as to whether it really is possible to rehabilitate and regenerate our historic towns, are awaited with growing interest in every country of Europe. *The techniques now being used in France will enable us to consider this problem in the correct context and in the correct proportions, for they have been specially evolved to serve the needs of whole urban centres and not only of individual architectural monuments.*

This brief survey of the historical development and the crystallization of the concepts 'architectural monument' and 'historic centre' does not by any means provide a comprehensive account of this subject. We have simply sketched out the main lines of this development in three different countries – Italy, Germany and France – which have made a special contribution to the problems involved and, by doing so, impressed on international opinion the need for protection. *Our principal object, however, is to furnish convincing evidence in support of the claims made in the introduction to this book, namely that 'the movement for the preservation of individual monuments was launched between 1820 and 1850, when the new industrial society was first beginning to assert itself, while the movement for the preservation of whole urban centres was started in our own day' – on the threshold of a new and foreseeable era of dynamic architecture and mobile planning.*

213   *Developments in Greece*

At this point, we propose to study the development of preservation in a small, relatively new European state, namely Greece, which possesses an extremely important artistic and historic heritage and is now faced with particularly difficult problems.

Ever since they acquired their independence in 1830, the Greeks have adopted a somewhat paradoxical attitude to historic monuments. They have always felt a profound respect for the cultural tradition of classical antiquity, and this has prompted numerous Greek scholars to take an active interest in archaeological research, in which they collaborated with many celebrated foreign scholars. Due to this widespread interest in antiquity, it was possible to excavate a considerable number of important monuments and cultural centres which had been buried for centuries, and to study them systematically. Many famous classical scholars took part in this archaeological campaign in search of antique art and culture, including Schliemann, Doerpfeld, Evans and Kavvadias. At the end of the nineteenth century, Byzantium was discovered. Diehl and Schlumberger and, subsequently, Vassiliev, Ostrogorsky and Runciman then studied and drew attention to the cultural traditions of the Eastern Empire, which had been unjustifiably neglected until then. Two Greek scholars, Sotiriou and Orlandos, devoted themselves to the study and conservation of the Byzantine monuments in Greece.

But, while they have certainly revered the works of antiquity, the modern Greeks have shown little respect for the living architectural monuments of the past five hundred years. Although there is no real justification for this paradoxical, indeed contradictory attitude, there are two explanations for it:

a) The affinity felt by the modern Greeks – and, for that matter, by all the peoples of the western world – with antique culture and Byzantine splendour stifled all interest in the more modest products of the five hundred years of Turkish rule, even though these possessed considerable aesthetic value and were of great importance for the cultural continuity of Greece. The architectural heritage of the Turkish epoch, like the nineteenth-century Greek Neo-classical buildings, have been systematically neglected, despite the fact that they constitute Greece's only living architectural monuments and her only inhabited historic centres. Apart from a few enlightened Greeks, who realized the importance of a living tradition, nobody appears to have been aware of their value.

b) The large-scale urbanization of Greece, which occurred about 1922, was accompanied by a spontaneous desire to make up for the stagnation which had been the hallmark of Greek architecture under Turkish rule. Although badly organized for the most part, the projects prompted by this activist spirit posed a great threat to the historic towns of Greece. These towns, the centres of which consisted of two or three-storeyed houses which were not protected by law against redevelopment, were the first to be demolished by the private builders, who had little or no idea of the cultural significance of this living heritage and had their sights firmly fixed on 'technological progress' and 'modernization'.

Like their Italian counterparts, the early Greek legislators were primarily interested in the protection of archaeological finds and sites. The Greek archaeological law of 1834 (which came on the statute book just four years after the declaration of independence) and the law of 1932 were both largely concerned with the penalties to be imposed for the illegal export of Greek antiquities. The new law passed in 1954 to protect 'historic sites' and the establishment ten years ago of a 'Department for Modern Greek Monuments' as part of the Greek Archaeological Service are the first really positive steps taken by the Greek administration for the protection of the country's living architectural heritage. Although quite inadequate, they do at least provide for legal protection and administrative control.

The Greek financial investment is also inadequate. The problem is not really understood, either by the public or – sad to relate – by the specialists. But then Greece has no real specialists in this sphere. Her architects have no means of pursuing the sort of studies available to the French 'custodians of monuments'. Consequently, the field is left to the archaeologists, who regard the living heritage as being of secondary importance and who are also inadequately informed about its problems.

But, although the administrative and legislative position still leaves much to be desired, individual voices have been raised since the founding of the new Greek state which have drawn public attention to what is at stake. A number of leading Greek scholars have called for greater respect for the living architectural heritage, and some have carried out extremely valuable scientific research in this field.

In this connection, reference should be made to Demetrius Cambouroglous, the Greek researcher and indefatiguable chronicler of the past five hundred years of Athenian history. He made the first inventory of Athenian monuments created under Turkish rule (1453–1827) and compiled the historical documents of this period. Cambouroglous was also the first scholar to insist upon the importance of the 'old' as opposed to the 'antique' aspects of the Greek cultural heritage – and he did so in the nineteenth century, when archaeological mania was at its height.

The architects Constantin Biris and John Travlos have continued in Cambouroglous's footsteps, albeit independently of one another. Both of these men have made searching studies of the city of Athens, taking account of the old town and its recent urban development (see Bibliography A and C).

Demetrius Pikionis, a talented architect and a professor at the Technical University in Athens, became one of the champions of the protectionist cause in Greece. With his lively mind and energetic personality, his expert knowledge of antique culture, his great admiration of Byzantine orthodoxy and his passionate love of Greece's living cultural heritage, he has won many disciples amongst the younger generation of architects and has also done much to influence public opinion.

The research work carried out before the Second World War by groups of students from the architectural faculty of the Technical University in Athens was issued in two well-documented and illustrated publications, which give an account of the patrician houses of Kastoria and Zagora, two historic centres in northern Greece.

Today other researchers – all of them architects – are doing similar work. Paul Mylonas, a professor at the School of Fine Arts in Athens, has been making a private study for the past ten years of the unique Byzantine and post-Byzantine architecture on Mt. Athos. During this time, George Moutsopoulos, professor at the architectural faculty of the University of Salonica, and a number of his students have continued the work started by Pikionis by investigating other historic settlements in northern Greece which have yielded important information about popular architecture. Such passionate and entirely voluntary involvement shows that the problem of protection is now fully understood in certain quarters in Greece.

The fate of the old town in Athens, the 'Plaka', has been the subject of lively debate of recent years, and the city authorities organized a public discussion of this question in 1966. The first indication that a serious attempt was to be made to rehabilitate the Plaka sector came when the Greek government instituted an enquiry, which was carried out by officials from the Greek Ministry of Works.

The protection and rehabilitation of the historic urban centres of Greece is now receiving more and more publicity, due largely to the initiative of the younger generation of architects, who have made a major contribution to this difficult problem.

11
Portrait of the Greek architect Demetrius Pikionis taken in 1967, a year before his death. Pikionis was responsible for the landscaping of the approaches to the Acropolis in Athens and of the neighbouring hills. Throughout his life he fought for the protection of Greece's living architectural heritage.

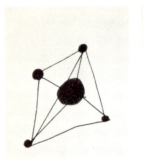

12
(Left) Diagram showing the disposition of major towns prior to 1900. These towns were a considerable distance apart and were connected by a single linear communications network, mainly traffic routes.
(Right) Diagram showing the disposition of major towns in the spatial cluster of the future. This arrangement is based on a continuum of closely interlocking networks with numerous centres situated very close to one another.

221   *The Urban and the Spatial Cluster*

We find historic urban centres both as independent units within, and as essential components (sectors) of larger urban complexes. In either case these centres constitute static and constant[10] urban formations within inhabited settlements built on the surface of the earth.

We refer to such settlements as general urban clusters (or artificial clusters) as distinct from the geophysical cluster constituted by the natural configuration and vegetation of the earth's surface.

The general urban cluster is composed on the one hand of a grid of linear connections (which carries the various networks), and on the other hand of 'centres' which are situated at the starting-points and inter-sections of these networks.

We now propose to investigate the different functions performed by the 'networks' and 'centres' within the traditional urban cluster.

The centres are urban formations which vary both in size and in functional importance[10]. They fulfil, either within a single nucleus (monocentric formations) or in a polycentric form (satellite towns and satel-lite suburbs), the functions of:

1. Residential centres;
2. Administrative, commercial and exchange centres (tertiary spheres);
3. Social, cultural, educational and health centres;
4. Industrial zones.

The traditional networks (i.e. the linear connections within the artificial cluster) fulfilled up to the be-ginning of this present century a function that was as old as the world itself, namely that of traffic routes. These networks consisted of roads, railways, rivers and sea routes.

The structure of the traditional artificial cluster as described above reveals the following characteristics:

a) *Its essential elements are static.* In the course of the past three centuries, there has been a marked increase in the number of urban centres and in the road, rail and water routes between them with the result that the density of the artificial cluster is now far greater than it was. The geographical location of these centres, however, and of the networks which link them has remained unchanged;

b) Due to the *isolation of the individual urban centres* and the largely inner-directed development to which this gave rise, their functional centres have been concentrated within the central urban area;

c) *There was no fully developed 'communications system' between the various urban centres.* The only 'communication' consisted in the traffic routes;

d) In the course of time, *the geophysical cluster was completely swamped by the artificial cluster.* The landscape was systematically ravaged and stifled.

From this it is abundantly clear that the traditional artificial cluster lacked 'continuity' and flexibility. Under this system, we find highly concentrated urban centres with inadequate inter-centre communications which stand out in sharp relief against the natural cluster that surrounds them.

The demographic explosion which has taken place during the past few decades and which will have doubled the world population by the end of the century, thus ensuring the widespread dissemination of industrial culture, has also contributed to the general increase in urbanization.

'Urban regions' (parent town plus outer suburbs plus satellite town or towns) have already been created on the eastern and western seaboards of the United States, in England, Belgium, the Ruhr, France and

---

10 The decadent condition of a few of these centres and the alterations made in certain networks are exceptions which prove the general rule.

Japan. These urban regions, incidentally, are the first stage in the transition from the traditional urban cluster to the spatial cluster of the future.

Constantin Doxiadis, the Greek town-planner and architect, has stated in various studies on the future development of urban formations that the exclusive characteristic of the future spatial cluster will be its extreme density. Such a disconcerting perspective is quite unacceptable, for it presupposes that the whole surface of the earth will eventually be covered by a massive, static artificial cluster that would throttle the natural geophysical cluster and so threaten the psychic and physical stability of the whole human race. However, we also possess other assessments of the future spatial cluster made by a number of progressive architects[11], who have posited the following characteristics:

a) *Extreme intensification of the linear connections between the various centres due to the development of new communications networks within the spatial cluster, which will ensure 'continuity' in the structured space of the future.* These new networks will consist of: 1. Telecommunications. (Today these include radio, telephone, telex and television; and these will soon be augmented by the television-phone and telemechanics [remote control of machinery], which will completely transform the tertiary sphere and revolutionize travel.) 2. Integrated networks for the distribution of energy (electricity and atomic power, solar rays etc.).

b) *The adaptability, flexibility and industrialization of architectural constructions.* The new architectural formations will make it possible for functional centres to be removed from the urban centres of the traditional artificial cluster and arranged in linear groupings of various kinds. These new centres will be based on the networks and consequently will be better able to cater for social requirements, and climatic conditions.

The traditional town, in which all the functional centres are concentrated in the congestion and confusion of the urban centre, will gradually be replaced by a new kind of settlement consisting of a continuum of different 'networks' (communications, telecommunications and energy distribution) and 'network centres' (residential, administrative, cultural, industrial) whose density can be modified at will by regrouping and which will therefore be eminently suited to the requirements of future developments within our urban settlements.

c) *The complete fusion of the natural geophysical cluster and the artificial spatial cluster.* Up to now clusters of human settlements have always been built above ground, thus changing and eventually destroying the landscape. In the future, however, part of these human clusters will be sited under the ground (subterranean urbanization for warehouses, factory buildings, car parks and link roads) while at the same time artificial platforms will be erected over rivers, lakes and bays.

But the greatest innovation within the future spatial cluster will still take place on the surface. Huge megastructures will be erected and will serve as bases for settlements, thus bringing about a flexible disposition of fully industrialized dwelling units, facilitating coexistence at different levels and ensuring the integration of landscape and settlement.

222   *The Significance and the Location of Historic Centres within the Urban and Spatial Cluster*

After surviving virtually unchanged[12] for hundreds and, indeed, thousands of years, the traditional urban centres expanded both upwards and outwards in the course of the nineteenth and twentieth centuries under the pressure of the radical urbanization process which is taking place all over the world.

In the majority of cases (for example in Rome and modern Athens), this new growth consists of concentric belts which have sprung up around the historic urban centre. In a number of cases, however, the develop-

11 In this connection see the writing of Paul Maymont, Yona Friedman, Paolo Soleri and the Japanese 'Metabolists'.

12 In his masterly study (*The Urban Development of Athens*, Athens, 1960) the Greek architect John Travlos has dealt in detail with the immutability of a major city.

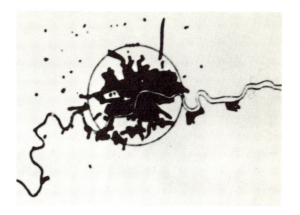
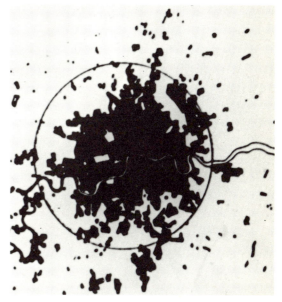
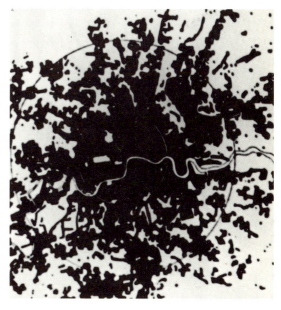

13
The urban explosion. Three phases in the expansion of London (1840–1900–1929).

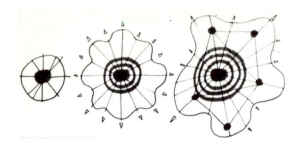

14
The radial-concentric growth of urban centres following the radial-concentric expansion of major towns and the establishment of subordinate centres on their outskirts.

ment has been asymmetrical and consists of new residential districts, most of which have been on the western side of the urban centre (as in nineteenth-century Berlin, Paris and London).

Expansion on such a scale inevitably caused a gradual breakdown in traffic flow and social functions within the urban cluster and eventually produced decentralizing tendencies. These first became apparent as early as 1920, when large residential and industrial settlements were erected on the outskirts of various major cities (e.g. Siemensstadt and Siedlung Britz in Berlin), which were largely independent since they had their own shopping centres and their own social amenities. This decentralization process became even more important after the Second World War when it produced both the satellite town (for example: Vällingby for Stockholm, Sarcelles and Massy-Antony for Paris) and the principle of planned polycentrism within major cities (on which the planning schemes for greater Paris during the past ten years have been based).

Apart from this polycentric conception, which has exerted what is virtually a 'de-urbanizing' effect and has certainly produced lower density levels, a further type of growth has been developed of recent years which is non-concentric and which anticipates the spatial cluster of the future. It is based on:

a) *The rectilinear or multilinear expansion of urban centres and of the functional centres which these contain (see 221)*;

b) *The structural alteration of the present self-contained centres and their transformation into a continuous network which would then constitute a spatial formation with an unlimited growth potential.*

In its initial phase, this kind of growth will be restricted to surface developments. Later, when the time is ripe for the full spatial cluster, these will be complemented by subterranean and overhead developments.

We shall deal on a later page with the role to be played by the historic centres as constituent elements of the future spatial image. Meanwhile, we must consider their function within the urban cluster of today and the spatial cluster of the future.

Within the framework of the traditional and static urban cluster, the historic centres (see 12) consist either of rural settlements, which have undergone little or no urban development, or of central suburbs in major cities. In either case they were doomed to gradual extinction. In the rural centres, this resulted from depopulation and the ensuing stagnation of social life, while in the central districts the principal cause of their decline was congestion.

Once the traditional structure of the historic centres had been deformed by the inappropriate schemes of the early developers – who widened roads to take motorized traffic and erected 'modern' buildings in a vain attempt to preserve these areas as centres of new metropolitan activity – they were gradually depopulated and slowly but surely fell into a decline.

15
The principle of free and linear urban expansion in one or more directions.

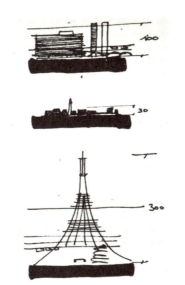

16

The three architectural scales within the urban cluster.

a) Present day. Constructions ranging from nought to one hundred metres.

b) The past. Constructions ranging from nought to thirty metres.

c) The future. Constructions ranging from nought to three hundred metres.

17

The close integration of the natural and the artificial cluster: a scheme proposed by Paul Maymont for a residential district to the west of Paris. In this plan towers, three hundred metres high, are connected by prestressed cables which are used to support prefabricated dwelling units. The various levels and areas are reached by ramps, lifts and overhead railways. Open areas, which are used for pedestrian precincts and community centres, are situated on artificial platforms fifty metres above the ground, which is given over to the natural cluster.

18

The magnificent countryside of the Swabian Alps in southern Germany showing the Hohenzollern Hill and the Royal Palace. In a hundred years from now, enormous structures could well bridge the valleys in this area, creating a spatial cluster in which the Hohenzollern Hill and the Royal Palace would serve as a visual point of orientation.

19

The present-day city centre of Athens with the Acropolis in the background and the Plaka, the old town of Athens, which has been left to decay on its northern slopes.

20

The town of Kastoria on a peninsula in western Macedonia. This small homogeneous provincial town contains interesting examples of popular architecture in the form of patrician and burghers' houses. After surviving unchanged for several centuries, these buildings have now been abandoned to their fate and have already reached an advanced stage of decay.

21

The small rural historic centre of Anticoli Corrado in the Sabine mountains of Italy. Because of its remote geographical position, this town has retained its close-knit urban structure throughout the centuries.

Eventually the town planners realized that, by trying to accommodate the tertiary functions of new and structurally alien centres within the old historic centres, they were simply destroying them. They then went over to the novel principle[13] of 'equal and parallel development', whereby the modern residential and social centres would be planned as a linear extension of the historic centres.

The adoption of this principle of linear expansion offers at last the possibility of protecting not only the historic centres of major cities, but all historic settlements.

Heretofore the historic centres have been used for administrative and commercial purposes within the traditional urban cluster. But they will stand no chance of survival within the spatial cluster of the future unless they are relieved of this role, which they are no longer capable of fulfilling within the modern social and urban structure. New (cultural) functions have to be introduced into the historic nuclei and at the same time their old (residential) functions must also be re-established (see 442) in order to ensure that life is preserved in the historic centres and their links with the modern urban and future spatial cluster are maintained.

We see, therefore, that the functional survival of the historic centres depends both on the adoption of linear rather than radial expansion and on the renewal of their earlier functions. This raises the question as to the kind of role to be played by our historic towns in the 'networks' and 'network centres' of the future spatial cluster.

13 The plan evolved over the past ten years for the urban development of Paris and the expansion of its historic centre constitutes a highly successful adaptation of this principle. The original proposals for this project, which were put forward in 1960 by a group acting under the aegis of the magazine *L'Architecture d'Aujourd'hui*, envisaged a new, parallel Paris, in which the centres of commerce and trade would be sited some sixty to eighty kilometres outside the historic centre of the city and would be linked to it by motorways and monorails. In other words, the historic centre was to have been reduced to a mummified monumental metropolis with no real life of its own. Fortunately, these proposals were rejected in favour of a more realistic and more effective approach. The present plan for the regional organization of greater Paris allows for the construction of several new centres on the outskirts of the urban area and for the expansion of the traditional centre of Paris westwards by extending the historic axis 'Louvre-Concorde-Etoile'. This vast project, which goes under the name of *Défense*, envisages two new purpose-built centres – one for commerce, the other for tertiary services – which will form a natural extension of the historic centre. Thus the new development will not be cut off from the historic areas of the city, but neither will it intrude into them. Looking into the future, the architect Paul Maymont envisages the incorporation of three hundred metres high megastructures on the same axis but extending further to the west and set in a landscape of artificial lakes flanked by the meanderings of the Seine.

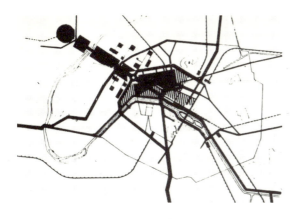

23

Model of the new Paris centre for the tertiary sector (*Défense* project) seen from the west. This development will form an extension to the historic axis Louvre–Concorde–Etoile.

22

The linear extension of the historic centre of Paris. The shaded areas represent the historic centre, the black areas represent the extension westwards, the first stage of which will consist of the *Défense* project, a large scale development scheme for a new residential and commercial centre. Approximate scale 1:200,000.

24

Street plan of the area to be used for the *Défense* project. The salient points of the new development are already present in this plan.

25

Aerial view of the historic East–West axis of Paris. In the foreground the Louvre and the Jardin des Tuileries. Behind them the Place de la Concorde, the axis of the Champs Elysées, the Arc de Triomphe and the Avenue de la Grande Armée. In the background – beyond the bridge of Neuilly – the new project will begin.

26 and 28
Scheme proposed by the Japanese architect, Kenzo Tange, for Tokyo (1960). Here Tange has taken the first step towards the spatial cluster of the future. His scheme envisages a settlement erected over Tokyo Bay by means of megastructures built on artificial islands. These megastructures, which would be up to thirty storeys high, would have two hyperbolic external surfaces, into which prefabricated dwellings would be incorporated. This new town would have vertical interior communications and would be linked with the outside world by clearways built at different heights and supported by suspension bridges. In this scheme, the natural cluster (sea and natural vegetation) would be closely integrated with the artificial cluster.

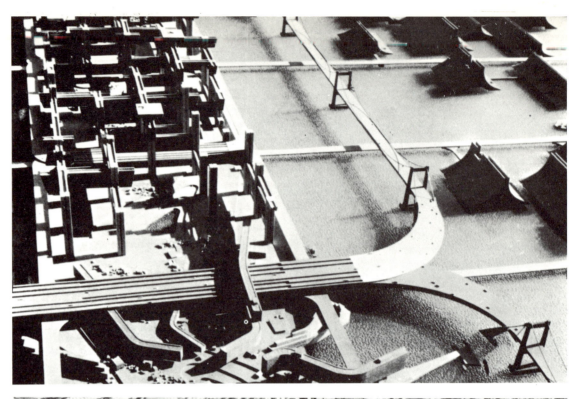

27 The endless expansion of a static traditional urban cluster built on the surface of the earth: a form found in England. Even when the houses are detached, as in the present case, this endless chain of buildings must inevitably destroy the natural cluster. Moreover, it then undermines the cohesion of the townscape and the dissolution of the urban settlement sets in.

Although our historic centres now cover only a relatively small area of the major cities to which they belong (in Paris *intra muros* 1/10, in Athens 1/100), their cultural and artistic value is very considerable. *Once their urban structure has been protected, renovated and rehabilitated, they will come to constitute within the future spatial cluster focal points of interest and culture which in the traditional urban cluster of today are represented by individual architectural monuments. The mobile architecture of the future with its completely new scale will use artistic and historical components as points of reference and orientation. In this sense the 'urban monument' will come to replace the 'architectural monument'.*

The only components of the future townscape which would be really suitable for such purposes will be the rehabilitated historic centres. In their new cultural capacity, they would provide a firm focal point around which the mobile and dynamic spatial formations of the future will be disposed.

After their functional restructuring (see 422, and their conversion into residential and cultural areas, the historic centres would also constitute a specific nucleus within the urban cluster, in which human life could follow a traditional pattern despite the advent of technology. This sphere would be given over to private life (residential quarters) and to the pursuit of intellectual, artistic and theoretical activities, for which man will presumably always want an intimate and specifically human environment, one from which the motor car will have been banished to make way for pedestrian precincts.

Over and above their symbolic importance as immovable monumental centres within the mobile and dynamic spatial cluster of the future, the historic settlements will also play a significant aesthetic and architectural role since they will provide an ideal setting for the noblest human activities both in private and in social life.

29
Perspective sketch executed by the author showing the sort of development likely to take place on the Plain of Argos in the Peloponnese a hundred years from now: megastructures will be mounted on gigantic pylons, which will cover the whole of the plain and serve as supports for residential cells. The natural cluster will remain unchanged, although the contours of the mountains and hills will be reflected by the line of the megastructures. Artificial islands designed like amphitheatres, which will provide accommodation for tourists, and small harbours for yachts will extend the artificial cluster out to sea. The rehabilitated historic town of Nauplia and the rehabilitated fortresses of Akronauplia and Palamidi will be integrated into this dynamic spatial cluster, where they will serve as visual points of orientation.

3   The Townscape

3 The Townscape

Historic urban centres are not a fortuitous conglomeration of monuments and other historic buildings. On the contrary, they are living urban districts with a specific structure which create a special kind of atmosphere.

In their capacity as 'living urban organisms', the historic urban centres are endowed with a quite specific 'townscape'[14].

31 CRITICAL SURVEY OF THE HISTORICAL DEVELOPMENT OF THE TOWNSCAPE

By analysing the different factors which have contributed to the formation of the townscape in the various epochs of European architecture, we shall be able to pinpoint those spatial principles which have proved to be the most important over the past two thousand five hundred years.

311 *The Feeling of Space in Architectural Creation*[15]

Various art historians and architectural theorists of repute[16] have rightly pointed to the difference between the feeling of space in antique architecture (eastern Mediterranean area and Mesopotamia) and western architecture (whose origins lay in ancient Rome).

The main characteristics of the antique architecture of the Near East, which reached the peak of its development in classical Greece, were the autonomous nature of individual buildings, which were erected in isolation within the urban space, the great importance attached to plasticity, and the invocation of absolute standards.

30

View of the sanctuary at Olympia from the southwestern entrance as reconstructed by C. Doxiadis. With its compact structure, which was based on its own architectural scale (the module), the Doric Temple of Zeus is typical of Greek antique architecture.

31

The well-lit nave and aisles of the Basilica of St. John the Evangelist in Ravenna. With its numerous windows, which serve to integrate the inner architectural space of the building with the outer space of the urban cluster, this basilica is representative of western architecture.

---

14 The specific townscape of the historic urban centre involves the 'inner townscape' (i.e. accumulation of perspective images and urban experiences inside the city) as distinct from the 'outer townscape', represented by the general urban cluster in its natural setting. A thorough knowledge of the concept of the 'townscape', of its historical development and the visual principles in which it is grounded, is essential if we are to understand the problems posed by the rehabilitation of historic urban centres.

15 We do not intend to furnish a systematic description of the historical development of European cities, for that would not be to our purpose. In this section we shall simply be referring to individual towns which illustrate various aspects of architectural theory.

16 See Giedion, [Space, Time and Architecture] *Raum, Zeit und Architektur*, Cambridge, Mas. and Michaelis, *Architecture as an Art*, Athens.

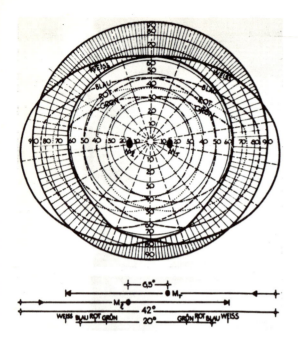

32
Binocular perception of space and colours. In the perception of form and volumes, the elliptical cone of vision has a vertical angle of about thirty degrees and a lateral angle of about forty degrees. Colour perception varies according to the colour. Green can be distinguished over a lateral range of ten degrees, while white registers over a range of twenty-five degrees.

In general, we may say that antique architecture created self-sufficient, three-dimensional forms, that it distinguished sharply between 'internal' architectural space and 'external' urban space, and that it was based on an internal or absolute scale or 'module' and completely ignored the human scale.

Western European architecture (which goes back to the massive vaulted constructions of the Romans, e.g. Roman baths and aquaducts) has developed along quite different lines. Here the principal characteristics have been the elaboration of 'internal' architectural space and the use of pierced volumes.

In general we may say that the architecture of western Europe is an architecture of internal space, that it links this internal architectural space with external urban space, and that it is based on an 'external' or human scale.

From this it is clear that the eastern architect of the antique world and his western counterpart had a completely different approach, and that in their basic conception of space they were literally poles apart. It has been suggested that these deep-rooted differences were due to racial characteristics. But the fact that the antique architects used different constructional techniques and that their buildings fulfilled very different functions would seem to offer a more plausible explanation.

In antiquity, the most important works of architecture (temples, graves, fortresses and palaces) were static structures built with 'beam and pillar'. Both their sacred and introverted function and their massive construction made for free-standing buildings with fully rounded and closed forms containing their own absolute or inner scale which was the module of the building (generally the lower diameter of the portico columns).

The great buildings of the western world (Christian shrines whose doors are open to everyone, civic buildings, baths, theatres, palaces), which began as vaulted structures and during the last two hundred years have been executed in steel and reinforced concrete, were conceived in terms of internal space. This was, of course, in keeping with their general function, which called for accessibility, and with their construction, which was not massive and which entailed the use of pierced volumes. Such buildings, which were based on an external or human scale, are not simply meant to be admired from afar. On the contrary, they invite the viewer to enter and study their interiors.

312   *The Feeling of Space in the Townscape*

The townscape in structured urban space[17] has not been determined by a unique spatial principle in any historical epoch. The character of the townscape results from the interaction of two principles, which constantly vie with one another for supremacy. These two principles, whose origins we shall try to establish, might be described as:

    a) The 'flat principle'.
    b) The principle of 'static, modelled form'.

---

17 By 'structured urban space', we mean the traditional urban space created by man through his arrangement of buildings and open areas within the urban cluster. Absolute space may be thought of as a transcendental continuum composed of innumerable units which, although differing in size, are nevertheless equal in value. 'Structured urban space' is one such unit. This relationship becomes clearer if we consider it in concrete terms: the difference in size between the internal space of a small Greek church on an Aegean island and the geographical space in which it is situated is enormous. But in terms of their essence, these two spatial units are nonetheless equivalent. Moreover, if we disregard the human scale, these two spatial units might also be held to be equal in volume: (in films we are often persuaded by means of trick photography to accept as real sea-battles waged in a bathtub). In our view, the distinction between absolute or cosmic space on the one hand and structured urban space on the other is far more meaningful than the traditional distinction between 'urban' and 'natural' space. For us urban and natural space are simply two of the innumerable and essentially identical spatial units which between them cover the whole range from the microcosm to the macrocosm.

To the human observer the townscape of structured urban space – like every other visual image – always appears as a flat, perspective image. In other words, it is a projection centred on the eye of the observer. This projection produces an elliptical cone with a vertical angle of about 30° and a lateral angle of about 40°. It also produces an imaginary plane of projection approximately 50 cm in front of the observer's eye which passes through the main axis of the optical cone at an angle of 90°.

The observer's angle of vision is widened by a second optical cone with the same centre (i.e. the observer's eye) and with a lateral angle of 60°. But in this outer space of vision, objects are not registered as firm forms with precise outlines: instead of an exact perspective image, the observer simply receives a general impression of colour and light.

Due to the 'binocular' nature of human vision, perspective images, which are actually flat, are enriched by an impression of depth. But this impression of depth, which is in fact an imperfect perception of physical volumes and the three-dimensional nature of space, is the only way in which man is partly able to grasp the essence of space.

Our imperfect perception of space is improved by a further factor, namely our direct experience of the urban scene. By 'physically' entering into urban space, we experience it in an extremely intimate way, thus acquiring an indelible impression and, as far as is humanly 'possible[18], gaining full access to urban formations.

From this it is clear that, although the townscape actually presents a 'flat' perspective image, the human observer is nonetheless able to perceive the three-dimensional nature of buildings and the 'spatial' character of their setting thanks to the psychological and physiological mechanisms outlined above.

If we now consider, in the light of these observations, the two opposing principles ('flat principle' and 'principle of static, modelled form') that are present in all townscapes, we find a ready explanation for their apparently paradoxical coexistence, namely the inherent duality of every perspective image.

The townscape vacillates between these two extremes. Depending on the historical period, it either appears to the human observer as a 'flat, perspective image' (in which case special importance is attached to the perspective view of the façades or the organization of the buildings on the linear axis) or else it produces an impression of depth, a plastic sense of static space[19], in which both the buildings and the open areas are autonomous elements. The strange thing – as we have already seen – is that this impression of plasticity can be conveyed by a flat perspective image.

33
The principal characteristic of urban formation in Greek antiquity was free alignment, which meant that buildings could be viewed from a wide variety of vantage points. Consequently, the observer was able to move through the urban cluster virtually at will and so 'experience' urban space. The antique Greek townscape reflected the plasticity of urban space. It was based on anthropocentrism, which the Greeks applied as a conscious principle.

18 It is a point worth noting that whereas human perception is subject to considerable limitations, certain species of insects possess quite extraordinary panoramic vision.

19 'Static space' is taken to mean traditional three-dimensional space, i.e. the structured urban space of our own times, as distinct from the 'dynamic space' of the future. Because of the inherent mobility and flexibility of its structures this 'dynamic space' will introduce the fourth dimension, namely time, into town planning.

34

The townscape presented by the triumphal routes and axially aligned Roman forums was quite different from the townscape of Greek antiquity. In Ancient Rome we find the prototype of the 'flat' townscape, which was based on symmetrical axes and the rhythmical alignment of colonnades. The dominant feature of this townscape was its frontal organization, which replaced plasticity as the fundamental principle of town-planning.

35

An example of the 'old' type of display in Greek antiquity (free composition built in stages). This illustration shows C. Doxiadis' reconstructed perspective view from the south-western entrance to the sanctuary at Olympia.

313   *The Townscape in Greek Antiquity*

The Ancient Greeks, whose most significant buildings were erected in accordance with the principle of free and asymmetrical urban organization, invariably created highly plastic townscapes for their important urban centres (sanctuaries and agoras). In this connection, it is a significant fact that even when completely different road systems were used (as in the case of the 'old' and the 'new or Hippodamian'[20] alignment), the design of the agora or sanctuary remained unchanged and always produced a powerful impression of depth.

In the 'old form' of road alignment we find terraced urban areas (e.g. the sanctuary of the Acropolis, the Athenian agora and the sanctuary in Olympia) in which the buildings are aligned in various directions. This 'free' organization was determined partly by the position of the temple (whose main entrance was invariably on its eastern side), partly by the need to guide visitors to the principal points within the urban cluster but, primarily, by the principle of free and asymmetrical disposition.

In the 'new or Hippodamian form' on the other hand, we find urban clusters (e.g. Priene and Miletus), which were conceived and built as corporate entities and in which both the buildings and the open areas were incorporated into the rectilinear grid of the road system.

20 Terms used by Aristotle in his *De Politica* to denote different road systems.

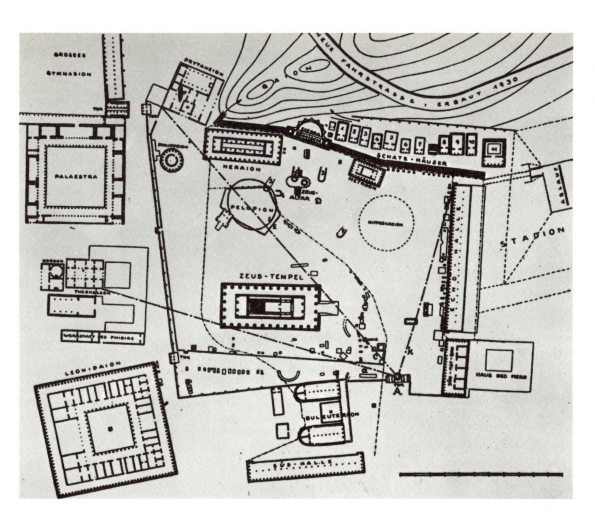

36
General plan of the sanctuary at Olympia as reconstructed by C. Doxiadis to illustrate his theory of the harmonic subdivision of the angle of vision into twelve equal parts. Approximate scale 1:250.

But even after the adoption of this 'new form', the sanctuaries and agoras, i.e. the most significant urban structures, still revealed no sign of axial or symmetrical organization. Almost invariably the perspective view was centred on a large open area that was cut off from its environment by a variety of clearly organized optical elements (buildings, high ground, perspective views of nearby natural features) and was built either on a single level (as in Olympia) or on various levels (as in Priene). Individual buildings of great symbolic significance, and structures such as altars or statues were often erected within the open area.

The buildings in Ancient Greek towns were never set out in line and *the perspective view of these free-standing structures created a townscape that was unmistakably stamped by the plasticity of static space and the latent dynamism of human movement*[21].

37
Bird's eyeview of the reconstructed model of the sanctuary at Olympia.

21 In the dissertation which he submitted to the Technical University, Berlin, in 1937, the Greek architect Constantin Doxiadis demonstrated the anthropocentric character of the urban formations of Ancient Greece by analysing the two major organizational factors on which they were based. These are:

a) the perspective view enjoyed by the observer from various critical points of access (propylaea) to the urban centres, and

b) the alignment of various roadways which suggested to the visitor the direction he should take within the urban space. These two organizational principles demonstrate the anthropocentric nature of the Greek townscape in antiquity. Incidentally, these principles were completely borne out by the thirty-eight investigations undertaken by the author in various antique centres in Greece. But a further conclusion of Doxiadis' theory, in which he proposed the harmonic subdivision of the angle of vision into ten or twelve equal parts and posited a subdivision of the visual rays based on the golden mean, is generally considered to be highly speculative.

Priene in Asia Minor: an example of the 'hippodamian pattern' (rectangular road alignment). Despite the steep terrain the freestanding buildings have none the less been set out in an asymmetrical pattern. By building the houses and squares at different levels and by erecting the fortifications in an irregular pattern, the original architect was able to introduce a strictly rectangular road network into this hilly site. Perspective views of the surrounding landscape are obtained from numerous points in the town with the result that the urban and natural clusters are closely integrated.

38 Perspective view of the agora in Priene with the Temple of Minerva in the background. (Reconstruction based on a sketch by C. Doxiadis.)

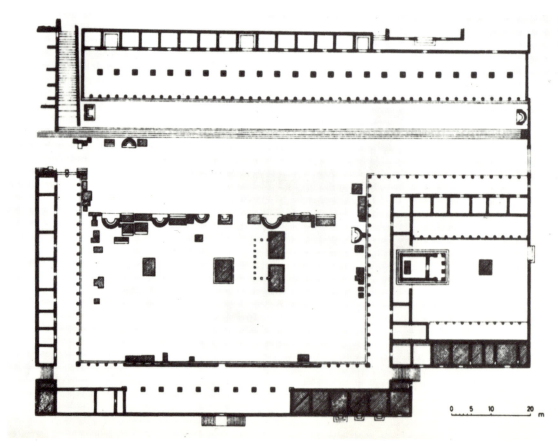

39 Ground-plan of the agora in Priene. Approximate Scale 1:500.

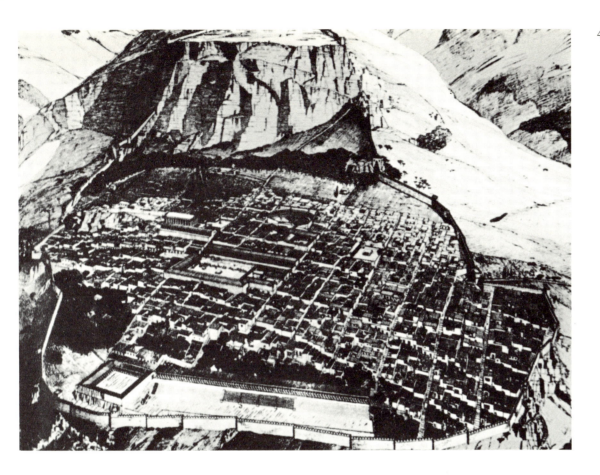

40 Perspective view of Priene and its surroundings.

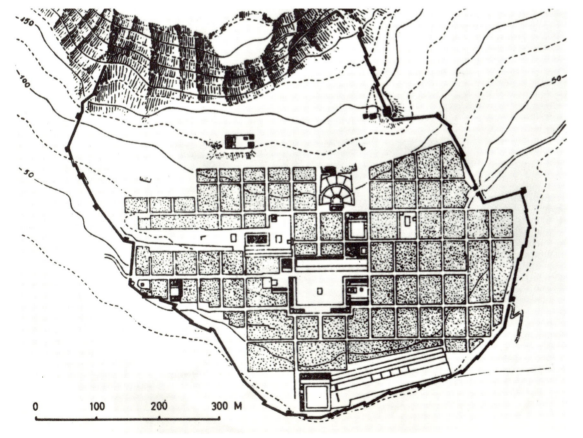

0   100   200   300 M

41 Street plan of Priene. Approximate scale 1:7,500.

42

Reproduction of the town of Halicarnassos in Asia Minor. General view from the sea. Note the remarkable integration of town and landscape: the clear-cut line of the mountains is repeated by the urban silhouette; the only building to rise above the general roof-line is the mausoleum, which serves as a monumental point of orientation.

43

The Stoa of Attalus II, King of Pergamon, which was reconstructed on its original site on the eastern border of the classical agora in Athens between 1956 and 1960. This Hellenistic portico with its rhythmically organized, close-set columns was one of the forerunners of the axial compositions (triumphal routes and 'flat' townscapes) of Hellenistic and Roman antiquity.

The Greek temples (which are generally regarded as the standard example of Ancient Greek architecture) were based on the module, the unit of length equal to the semi-diameter of a column at its base, which represented the absolute and inherent mass of a given building as an abstract and plastic work of architecture. Consequently these temples completely ignored the human scale. This was not the case, however, with the urban compositions of Ancient Greece, which were attuned to the latent and apparent movements of man and created a townscape that took account of the most important sites within the urban cluster and of the principal vantage points accessible to the human observer.

Symmetrical and axial town plans which create a 'flat' townscape were completely alien to the early Greeks. Their urban centres always had openings and perspective views which enabled them to incorporate the natural environment into their urban compositions.

The same spatial principles – i.e. the application of a human scale and the close integration of urban and natural space – were also characteristic of the outer urban areas of Greek settlements. Greek towns were seldom built on plains and were never very big. The nature of the terrain (hills, slopes) greatly influenced the general shape of their urban structures[22].

The principal characteristic of the 'outer' townscape of the Ancient Greek urban cluster (which was conceived as a free and asymmetrical formation on both the horizontal and the vertical plane) was its integration with the surrounding landscape.

314   *'Flat' Townscapes in Antiquity*

The 'planar principle' in antique town-planning, which produced 'flat' townscapes, is first seen in the perspective view of the close-set Hellenic colonnades or stoas, which were relatively late creations. Although these colonnades were found only in individual buildings and so did not constitute a whole urban image, they were nonetheless direct precursors of the first axial compositions (monumental triumphal routes) of the Hellenistic and Roman periods, which were the earliest urban formations presenting 'flat' townscapes in the history of European town planning.

The Romans were not the first to develop this highly symbolic system of axial routes. The Ancient Egyptians had anticipated them long before, although in their case the routes had not been fully developed and consequently had not provided a real 'image of the road'. In fact, they had been purely symbolic and plastic alignments (bilateral alignment of the sphinx etc.) which had merely indicated a rectilinear approach to the temples.

The roads in the Greek sanctuaries and agoras were also purely functional. Although the most important buildings within these precincts were linked by roads, no attempt was made to provide these with a façade. Even the roads in the Greek residential areas were no more than traffic routes for, since the buildings all faced away from the roads, these did not present an 'image of the road' in any real sense, but simply a collection of blank walls.

The Hellenistic and Roman roads, many of which were more than twenty metres wide, were entirely different. With their colonnades and triumphal arches they provided a genuine 'image' and so constituted the first fully developed example of a symmetrical alignment built on a linear axis. *During this period roads ceased to be mere traffic routes and became a visual component of structured urban space.*

---

22 The virtually systematic rejection of level sites in favour of hilly terrain cannot have been entirely due to the need to take protective measures against invaders. The towns of the Roman empire, which were constantly attacked by barbarian hordes, were almost invariably built on level ground. In our opinion this happy choice was very probably due to the existence in Ancient Greece of a highly developed collective sense of spatial values.

In the axial system the roads were not flanked by a loose arrangement of free-standing buildings, but by regular façades running parallel to the central axis. Although the rhythmical repetition of the columns of the portico and the perspective view of the horizontal roof-lines of the buildings created a certain impression of depth, they did not produce a sensation of three-dimensional space in the full sense of the word. The rich and varied spatial experiences, which are possible when man enjoys freedom of movement in urban space, are precluded by the axial system, which imposes limitations on movement by the rigidity of its alignments.

The perspective view of the two parallel rows of façades produces a sense of profound monotony due to the repetitive nature of the architecture. In place of the spontaneity and the unexpected views which are the characteristics of free alignment and free-standing buildings, the axial system creates an aura of rigid grandeur.

The focal point of the urban composition is no longer the mobile human observer but the symmetrical axis and the linear organization of the road system. Man's feeling of space is undermined for despite the emphasis laid on perspective views, the townscape no longer represents an organized, three-dimensional urban formation but simply an accumulation of flat façades. It is, in fact, the prototype of the 'flat' townscape of our European settlements.

The anthropocentric principle, which determined the structure of the Greek agora, did not play a major part in the design of the Roman forum. The forums were not open spaces but enormous rectangular or quadrangular courtyards with colonnades on all four sides. Like many other Roman urban areas, they were built on a linear axis. With their perspective view of the colonnades and their axial organization, they too provided a townscape that gave little real impression of depth.

A further result of the 'closed' character of Roman town planning was the complete absence of perspective views of the surrounding countryside. The failure to integrate the landscape into the urban pattern (which was due in part to the flatness of the terrain) produced in the majority of cases an 'outer' townscape that was extremely monotonous. In Rome itself, as in most other Roman towns, the urban structure was made up of a wide variety of highly disparate complexes which had been built as the need arose.

315   *The Townscape of the Medieval Settlements*

Because of their social organization and urban structures, both the Byzantine and the West European medieval states had far more in common with Ancient Greece than with Imperial Rome – and this despite their view of the world, which had no counterpart in antiquity.

With the single exception of Constantinople, which modelled itself on the Roman metropolis, the medieval towns of eastern and western Europe developed either as regional, feudal capitals (France and Byzantium) or as trading towns (Italy, Flanders and Germany), both of which had more in common with the settlements of Ancient Greece than with those of the Roman empire.

The Romans, whose cities formed part of a uniform network of provincial centres, built on a massive scale. By contrast, both the Ancient Greeks and the people of the Middle Ages, whose independent towns and cities were designed for a limited population, built on a far more personal scale[23].

44
Plan of the Forum Romanum (Forum Trajani and Forum Augusti) in Rome. The enormous squares with their interior colonnades, which were displayed on a longitudinal axis, created a completely scenographic townscape with little or no plasticity. Approximate scale 1:5,000.

45
Detail of the reconstructed Forum Romanum.

46
Perspective aerial view of Aigues-Mortes in the south of France, a medieval town that was built as a rallying-point for the Crusades and shows a rectangular road network. Approximate size of the settlement: 530 × 290 metres.

23 This also applies to the Byzantine empire, where – apart from Constantinople (the new Rome) – no new metropolis was built. The Roman empire proper, on the other hand, boasted a large number of major cities including Carthage, Alexandria, Tyre, Antioch, Pergamon and Corinth.

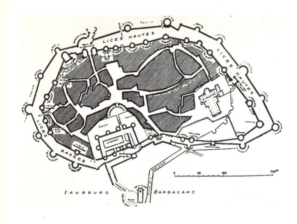

47

Plan of Carcassonne in the south of France. Approximate scale 1:8,000.

The larger medieval towns, which evolved over a long period, often enjoyed the protection of feudal institutions (castles, monasteries). The smaller rural settlements, which were situated in isolated areas, were provided with an ingenious defensive system by the simple expedient of strengthening the outer walls of the houses on the periphery of the town. Thus enclosures were created without actually building town walls, which were found only in larger towns.

The layout of the streets in such towns was usually quite irregular, although in the late Middle Ages (fourteenth century) there was an interesting development, which constituted a marked departure from this general rule: a number of small provincial centres (e.g. the Bastides in France and Terrae Muratae in Italy) were built with a completely rectangular street plan. This was a direct consequence of the social changes which were then taking place. In Italy the middle classes were becoming more and more powerful, while in France the centralized monarchy was engaged in a bitter struggle with the feudal lords. It was these new forces that advocated the establishment of small urban centres (with an overall diameter of between two and six hundred metres and with a strong defensive wall) in which their followers might find protection from the feudal aristocracy.

The inner townscape of the medieval urban centres reveals a degree of plasticity comparable in its own way to that found in the settlements of Ancient Greece. But this new plasticity was created in a new way. Instead of erecting free-standing buildings like the Ancient Greeks, the medieval architects created compact blocks of buildings with irregular roof-lines which they interspersed with buildings of special symbolic and practical significance such as cathedral watch-towers and town halls. The variations of height and mass between the different houses and the towering bulk of the public buildings produced innumerable focal points of interest to engage the attention of the observer. The most important edifices were graced by squares, esplanades and courtyards, which were often broken up into two or more disparate sections by the encroachment of other asymmetrically sited buildings, thus affording interesting and beautiful views of inner urban space and of the surrounding countryside.

The new system of vaults and their pierced supporting surfaces, coupled with the towering structures of medieval architecture (which emphasized the vertical aspect of urban composition), produced a new form of plasticity that was comparable to that of ancient Greek formations.

To realize the visual superiority of medieval urban compositions over those of other epochs, we need only consider the psychological 'urban' experiences which they offer[24].

If we walk through a medieval town we find something new at every turn. The freely aligned roads lead us to the focal points of secular and ecclesiastical life; the squares – also freely aligned and more or less enclosed – set up an interrelated pattern, while the roof-line rises up from the human level of the vaulted colonnades to the proud towers and fortresses and, finally, to the lofty spires and cathedral naves or – in the Byzantine settlements – the broad, soaring cupolas of the great churches. All these elements go to make up the great wealth of urban experiences offered by the medieval town.

In the urban setting, which creates an intense impression of three-dimensional space, the 'townscape' never appears to be 'flat'. In this it differs from the urban image of Imperial Rome, of the Baroque and of the nineteenth century. Axial scenographic alignment was entirely alien to the Middle Ages, just as it was to Ancient Greece.

The outer townscape of the medieval town was governed by the same principles as those applied in Ancient Greece, which meant that the urban cluster was integrated with the surrounding countryside. This integration was due partly to the fact that different parts of the town were built at different levels and partly

---

24 It is important to realize that our comprehension of urban space depends to a considerable extent on the psychological mechanisms involved in our 'urban experiences'. It is these experiences which enable us to grasp human and architectural factors and to recognize the wide range of movements and relationships available to the inhabitants of urban areas. Human perception undoubtedly depends on a combination of physiological and psychological factors: 'ΝΟΥΣ ΟΡΑ ΚΑΙ ΝΟΥΣ ΑΚΟΥΕΙ' (The Mind sees and the Mind hears).

48

Aerial view of Carcassonne. The attractive outer townscape of this well-preserved medieval town derives from the successful integration of the urban cluster with the surrounding landscape. Although the massive fortifications isolate the town from the natural cluster on the lateral plane, this isolation is relieved by the defensive wall, which follows the natural line of the terrain, and by the presence of various vertical elements (towers and belfries), whose soaring forms link the urban cluster to the atmosphere.

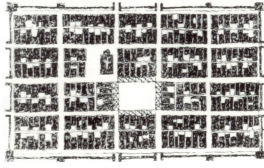

49

Plan of Monpazier in France. This was one of the few medieval towns to have a rectangular road network. Approximate scale 1:5,000.

50 and 51

Aerial view and plan of the town of Nördlingen in southern Germany. A typical example of a medieval urban cluster with a high density of urbanization and free road alignment. Although this aerial view reveals a circular and star-shaped pattern of street networks, this is not recognizable as such from the ground.

Aerial view and plan of one of the most interesting extant medieval urban clusters: the centre of the old town of Zürich with the Grossmünster (1260) and the Frauenmünster (1150–1400) on either side of the Limmat river. The marked plasticity of the townscape of this settlement derives from the typically medieval arrangement of the buildings on different levels, from the free alignment of the squares and their subdivision into component areas of varying sizes (due to the asymmetrical arrangement of the churches and other important buildings), and from the dense network of narrow streets with their highly disparate and irregular buildings. Approximate scale 1:5,000.

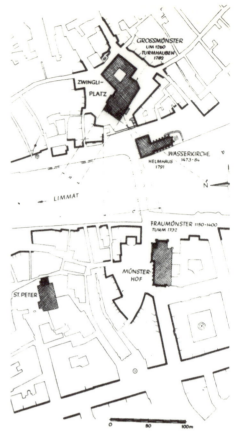

to the great disparity in size between so many of the buildings. Although medieval towns were cut off from the open countryside by fortifications, they were not alienated from it. What they lost in lateral integration, they more than made up for in vertical integration. Their soaring structures, which were in perfect keeping with their natural setting, demonstrate yet again the visual principles on which these towns were based and which produced these almost perfect examples of 'unity in multiplicity'.

In the Middle Ages we also find a small number of 'flat' townscapes. These few exceptions to the general rule would bear out our thesis that the two fundamental principles in urban composition (i.e. the 'flat principle' and the principle of 'static, modelled form') are present in every period of culture. These exceptions may be placed in two principal groups:

a) 'Flat' street façades in medieval towns which were developed either on the coast or on the banks of rivers. In this particular case the 'flat' townscape will have been due to topographical factors rather than to a conscious desire to create a geometrical and frontal urban composition.

b) Temporary formations of a special kind, namely movable military camps which were built in the midst of open country. The architectural structures in these camps consisted of tents, sunroofs, palisades, and flagpoles and may, therefore, be regarded as spatial clusters *sui generis*. Such formations were prompted by practical and strategic considerations and by the desire for ostentatious display. Consequently, they did not have a social life in the true sense of the word. It was this combination of motives that led to the development of the first geometrical urban formations, which were the precursors of the ideal cities of the Renaissance. Because of the rhythmical organization of their component elements and the use of the three basic Gothic colours (red, yellow and blue), these formations created the first 'flat' and virtually scenographic townscapes to be found in the Middle Ages.

Before going on to consider the Renaissance and the Baroque townscapes, let us first recapitulate the most important points in our critical survey of the evolution of the urban image so far.

In antiquity the Greeks evolved humanist, aristocratic[25] and more or less democratic social structures for which man was invariably the point of departure. *With the Ancient Greeks anthropocentrism was not fortuitous but a fully conscious principle.*

In the autocratic Roman empire, on the other hand, man was regarded as a subject of the state rather than as an individual. Consequently, instead of producing anthropocentric urban clusters, the Romans created rigid structures based on axial alignment, in which man played the passive and subservient role of a spectator. *In Ancient Rome the humanist ideal was replaced by the totalitarian principle.*

The Middle Ages, in which people lived in close-knit communities under the shadow of the Church and the feudal aristocracy, produced their own special brand of humanism. It was a long time before this was generally appreciated for it was felt that the theocratic character of medieval secular power and the feudal structure of medieval society were incompatible with the humanist ideal. In point of fact, however, the urban centres of the Middle Ages (most of which are still inhabited) reveal structural qualities which testify to a new kind of humanist approach: instead of being the point of departure for urban composition, medieval man became its mobile and omnipresent point of reference. *Consequently, the anthropocentrism of the Middle Ages was not an intellectual principle, as it was for the Ancient Greeks, but the product of daily urban experience.*

After the Middle Ages, the Italian Renaissance and, subsequently, the French Baroque introduced into the social life and philosophical outlook of western Europe a new humanist conception, which is still the main driving force of western society today. Because of the rise of the *bourgeoisie*, the formation of national states and the growth of absolutism, man was regarded neither as a point of departure nor as a point of reference but as the fixed geometrical centre of urban composition. *This led to an exaggerated form of anthropocentrism based on an over-assessment of the importance of the individual.*

Between the fifteenth and nineteenth centuries, the European townscape was determined by the axial and symmetrical character of urban formation. With its emphasis on ostentatious display this new western approach created a completely 'flat' townscape, which was brought about by a general reduction of plasticity within urban space and the concomitant growth of 'façade architecture'.

The Renaissance saw the large-scale development of axial and 'central' compositions with their visual centre or focal point based on this 'principle of urban organization'. We find this conception in individual buildings (e.g. Palladio's Villa Rotonda in Vicenza), in small urban compositions (squares centred on vertical axes represented by obelisks etc. such as the Piazza del Popolo in Rome) and also in composite urban centres (e.g. in the celebrated 'ideal cities'[26]).

The Baroque architects were less concerned with the symbolic value of the 'centre' and tended to attach greater importance to the axial character of urban formation. Although a number of important urban compositions were still created around a central point during the seventeenth and eighteenth centuries, this point was not emphasized by means of visual structures (obelisks, fountains). Two outstanding examples of this type of design are 'La Place Royale' (Place des Vosges) in the Paris district of the Marais and the quadrangular inner courtyard of the Louvre.

---

25 Aristocratic in the Aristotelian sense of 'the power of the best' (see Aristotle, *Constitution of Athens*).
26 Although many such cities were planned and aroused great enthusiasm, only very few were built (Palmanova, Freudenstadt and Vitry-le-François are examples). But the plans, which have survived and which were based on circular, polygonal and star-shaped forms, clearly illustrate the principle of 'central' composition.

54
Aerial view of Palmanova, one of the ideal towns of the Renaissance, which was founded in 1593 in the vicinity of Udine in Italy. The original plans were drawn up by Scamozzi.

55
The Place Royale (now the Place des Vosges) in the historic urban sector of the Marais in Paris. In its original form this square, which dates from the French Late Renaissance, was a splendid example of central composition. Unfortunately, the French nineteenth-century town-planners changed its character by planting trees in order to bring it into line with their own romantic notions.

56
St. Peter's Square in Rome. The quintessence of central urban composition arranged around a vertical axis (obelisk).

57
The Place Vendôme in Paris. The strictly symmetrical façades of the buildings (façades ordonnancées) are grouped around the famous column of the Grande Armée, which serves as a central feature.

In the urban compositions of the Renaissance, the Baroque and the Neo-classical Napoleonic era, individual 'noble buildings' played an important part. With their interrelated wings and courtyards these monumental edifices constituted independent urban formations within the framework of the general urban cluster. Many such buildings were created in Paris and the French provincial towns (e.g. Bordeaux and Nancy), in Austria, Germany and Spain. They were situated either in the centre of the town or on one of its linear extensions, and consisted for the most part of royal and princely palaces (Versailles, Louvre, Nymphenburg, Schönbrunn) or important administrative buildings. The image of these urban formations, which was invariably flat and virtually scenographic, set the pattern for the development of the feeling of space in the nineteenth-century urban cluster.

The seventeenth and eighteenth centuries saw the emergence of long and ordered façades (façades ordonnancées). By the terms of a French royal decree, monumental buildings erected on streets and squares had to have identical façades, irrespective of their owners' wishes: all colonnades had to be of the same height, the cornices had to be continuous, the buildings had to have the same number of storeys and the same type of roof. The Place Vendôme (1686–99; architect Jules Hardouin Mansart) and the Rue de Rivoli (laid out in 1806; architects Percier and Fontaine), which are two examples of government intervention into the 'untouchable' domain of private property, still impress us by the harmonious proportions of their general design.

The 'flat' frontal character of European urban formations has often been achieved by the deformation of colonnades, i.e. by using colonnades, not as functional components of buildings, but as purely decorative elements in the creation of a street façade.

During both the Baroque and the Rococo periods, this decorative trend was sometimes carried to such extremes that urban space was transformed into what were virtually stage sets. One example of this development is the Piazza San Ignacio in Rome. Here the ornate façades and the completely incongruous angled corners of the circular group of residential buildings combine with the deceptive impression of depth and perspective created by the roads converging on the piazza to form the *non plus ultra* of the scenographic urban image.

The Castle of Chantilly, one of the most important castles in the vicinity of Paris.

59

The Nymphenburg Palace in Munich. The open areas surrounding this castle – like those at Chantilly – are set out in geometric patterns. They have, in fact, been completely 'architecturalized'. No attempt was made to integrate the buildings into these open areas. On the contrary, the architecture and the landscape design – both of which are based on geometric principles – were brought into open and monumental confrontation.

60

The Bois de Boulogne, now a large park to the west of Paris, prior to its restructuring under Napoleon III: woodland covering 850 hectares. Approximate scale 1:50,000.

61

The Bois de Boulogne after its restructuring during the Second Empire (1852–70). The original forest was transformed into a landscape park with winding paths and artificial lakes. This romantic composition was based on English models. Approximate scale 1:50,000.

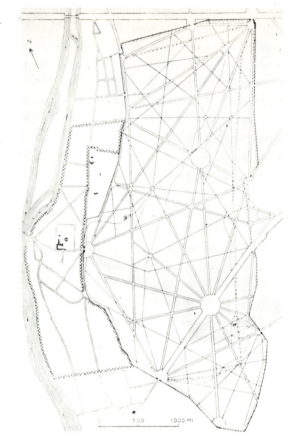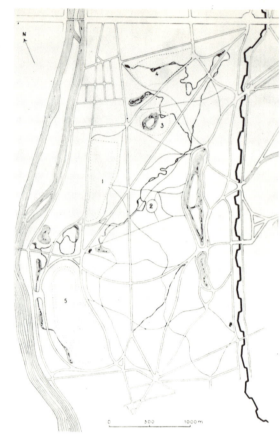

62

The outer courtyard of the Louvre and the Jardin des Tuileries in Paris. This is a further example of a large open area in a major European city with a strictly geometric layout.

As for the importance of architectural complexes during this period, the fact of the matter is that whereas the Renaissance produced a number of entire urban centres (e.g. Florence, Bologna and a significant part of Rome), the Baroque concentrated almost exclusively on monumental palaces, squares and urban axes.

In the Baroque the integration of the urban and natural clusters was replaced by the 'architecturization' of nature. The grandiose geometrical schemata of this period reveal axial and central structures which appear to transcend the architectural organization of the urban cluster, thus encroaching on to the natural cluster. The Baroque parks and gardens, which were conceived both as decorative features and for purposes of display, formed an integral part of the general architectural composition.

For the modern observer there is one element in this kind of composition which creates a three-dimensional effect and produces a feeling of space, namely the arrangement of steps, ramps, cascades and fountains. In this connection, however, we must bear in mind that modern man with his aesthetic background is particularly receptive to three-dimensional effects and many well notice features in these gardens to which little attention was paid during the Baroque period. We consider, therefore, that originally these waterfalls and steps were intended to enhance the completely scenographic character of the gardens from various important vantage points[27].

27 This interpretation is borne out by the extant etchings of the period, which invariably present a general axial view of the gardens. There are no contemporary etchings showing perspective views from closer vantage points, which would have drawn attention to the differentiation produced by the various levels and to the interplay of plastic elements.

A view of the arcades on the Rue de Rivoli in Paris. With their ordered façades – which were prescribed under the terms of an imperial edict – these arcades are typical of the nineteenth-century 'flat' townscape. The ostentatious architecture of the period reached its peak in the rhythmical repetition of the pilasters, lanterns and windows and the pronounced line of the cornices on these buildings.

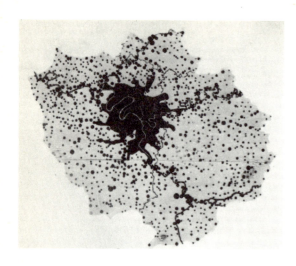

64
The demographic explosion in the greater Paris area, which now has a population of over nine million. The central section (within the white contour) represents the 'historic Paris', which occupies approximately one-tenth of the greater urban area and houses almost half the population. Approximate scale 1:2,000,000.

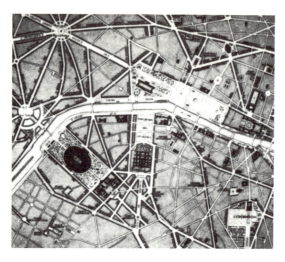

65
Plan of the interconnecting boulevards and avenues designed by Haussmann for Paris. Most of these were built during the Second Empire. Approximate scale 1:50,000.

66
An example of monumental radial alignment: the Place de l'Etoile (diameter 250 metres) with the Arc de Triomphe (erected in honour of Napoleon I).

317   *The Nineteenth and Twentieth Centuries*

The nineteenth century, in which the first industrial revolution took place and the *bourgeoisie* finally established itself, also saw the development of the new technological phase of architecture and the emergence of the first major cities with populations of over one million. The concentration of large numbers of people in city centres and the availability of new building techniques led to brutal changes in the character of urban space. The townscape suddenly assumed new dimensions.

But, despite these new factors, the traditional style of architecture and the established character of the townscape scarcely changed. On the contrary, architecture entered on a stagnant phase, in which it was completely preoccupied with imitative works.

Eclecticism in architecture is never fortuitous. The triumphant bourgeoisie of nineteenth-century Europe celebrated its emancipation from its former subservience by going through the whole gamut of traditional architectural styles. This is why neo-Classical, neo-Gothic, neo-Renaissance and neo-Baroque architecture was promoted in such rapid succession during the second half of the nineteenth century. With their development of 'façade architecture' (in which the façade often played no part in the functional organization of the building) and their insistence on display, the architects of nineteenth-century Europe finally reached the heights of absurdity.

The major innovation within the sphere of nineteenth-century town-planning is to be found in the new roads and avenues, which were built from 1850 onwards and which reveal the most important characteristics of the 'flat' townscape: linear arrangement of trees on the avenues and highly ornamental façades of buildings with identical roof-lines.

These features typified the townscape of the nineteenth-century European metropolis which – despite the technological, functional and demographic changes that had taken place – was still inspired by the same insistence on display that had been the quintessential quality of seventeenth-century Baroque and consequently still revealed the same 'flat' character.

But while urban architecture was wilting under the eclecticism demanded by an ostentatious society, the landscape gardeners were creating highly aesthetic open air parks – a contrary trend which spread from England to practically the whole of Europe in the course of the nineteenth century and was based on the natural landscape of northern Europe. By digging artificial ponds, lakes and canals, by planting clumps of trees and laying out meadows within a natural setting, these landscape architects created the only works of their era to reveal a genuine sense of three-dimensional space.

During the first half of the twentieth century, 'modern' architecture succeeded in banishing the evil of eclecticism. Both the new residential areas built in the environs of our major cities and restructured districts[28] produced a genuine feeling of space.

This 'modern' conception of the organization of urban space, which was based on functionalism and a new kind of social humanism, was supported by Le Corbusier (in his *Urbanisme* of 1924) and was expounded in the 'Charter of Athens'. The new movement succeeded in reintegrating urban and natural space and also re-established the supremacy of 'the feeling of three-dimensional space' in the sphere of town-planning.

The dominant feature of the modern townscape was its total rejection of nineteenth-century 'façade architecture' with its rows of identical façades running the whole length of the street.

But, although the 'flat' townscape was forced to yield pride of place for the first time in four hundred years, it did not disappear from the scene altogether.

In fact, the chief characteristic of urban development in the first half of the twentieth century has been the open co-existence of the two fundamental aesthetic principles ('flat principle' and principle of 'modelled form'), which have always run parallel to one another but which have never before been placed in such sharp relief. On the one hand we find compositions, whose plasticity derives from the disposition of independent buildings on open spaces within the urban cluster and which have come to symbolize the new aesthetic and functional doctrine of architecture, while on the other hand we find that the 'flat' townscape has also put forth new shoots.

The façade architecture of the nineteenth century, which was completely indifferent to the internal functional requirements of buildings, has been revived in the 'curtain walls' of the present period. These entirely simple and highly imposing surfaces of glass form an outer shell for buildings, which is not connected with inner functions and structure. It is important to realize that what we are faced with here is a 'modern' form of 'flat' townscape. But there has also been another development in the first half of the twentieth century which constitutes a marked departure from the 'plasticity' of the modern townscape.

This is the disposition of high buildings (tower blocks) in rectangular or linear series (i.e. rhythmical repetition. Le Corbusier's 'Voisin' plan for Paris is an example of the first development and the high buildings in the new city centre in Stockholm are an example of the second.) These axial and – in many cases – symmetrical compositions can scarcely be reconciled with the principle of plasticity in groups of freely organized buildings.

During the past twenty years, however, an attempt has been made to do away with this rhythmical organization by creating a townscape which, although still static (non-mobile), nonetheless produces a really new kind of plasticity.

67
The longest linear axis in Europe: The East-West axis in Berlin, which starts in the traditional urban centre as Unter den Linden and measures twelve kilometres from end to end. This photograph, which was taken before the Second World War, shows the Brandenburg Gate with the Victory Column in the background.

---

28 Such districts are to be found in cities like Rotterdam and Berlin, which were devastated in the Second World War, and also in places like Stockholm, where the city centre was restructured as a matter of policy.

68

a) Nineteenth-century urban composition: identical blocks built as 'closed formations'.
b) Twentieth-century urban composition: buildings of different sizes freely aligned.
c) Residential blocks in Le Havre (France) before the Second World War.
d) The centre of Le Havre following its radical restructuring and reconstruction.

69

Le Corbusier's proposed scheme (the Plan Voisin, 1922) for the restructuring of the centre of Paris, which was – and still is – a source of heated controversy. These free-standing tower blocks of uniform height do not produce the sense of plasticity that is the principal feature of urban composition today.

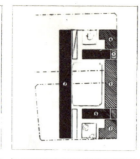
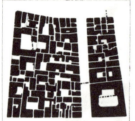

70

Tower blocks near London. Here too the disposition of the blocks lacks plasticity.

71

The monumental administrative centre of Brasilia, the most successful example to date of the new sense of plasticity in urban space, which was developed in the second half of the twentieth century and has introduced a new kind of diversity into the dynamic integration of buildings and open areas.

A significant example of this new development is afforded by the Square of the Three Powers in Brasilia. Here a number of free-standing buildings, which differ both in form and in size, are grouped around the main axis of the urban composition where, by virtue of their complete asymmetry, they produce a new feeling of harmony. Due to their close integration with the large courtyards and squares, by which they are surrounded and which are themselves closely integrated with the natural setting, these buildings constitute the most successful attempt to date to put the new three-dimensional conception of urban space into practice. Because of its significant dimensions, the clarity of its geometrical forms and volumes, the latent dynamism of its asymmetrical organization and the close integration which it proposes between the urban and the natural cluster, this conception must be regarded as a direct precursor of the future spatial cluster, which has already been envisaged by visionary architects.

321   *The Townscape as a Visual Composition*

In the light of the preceding critical analysis of the historical development of the townscape, we can now advance the following definition:
The townscape is the visual and aesthetic aspect of an urban composition.
This man-made composition has numerous constituent elements, all of which have specific functions (residential, educational and work spheres; traffic, energy and telecommunications networks; public areas; green belt areas etc.). But these functions exert no visual effect and so play no part in the formation of the townscape, which derives solely from the *visual* organization of the constituent elements within the urban cluster.

322   *The Townscape as a Spatial Composition*

This kind of townscape reflects the reality of three-dimensional space. The townscapes of urban centres (even those which have acquired a scenographic character in the course of recent centuries and so present a 'flat' frontal or perspective image) invariably represent a specific conception of the organization of structured urban space.
Until recently this organization had always been conceived within the static framework of the three spatial dimensions and in its application had always presupposed permanent contact with the surface of the earth. Its most important characteristic was the rhythmical interplay or the harmonic and asymmetrical arrangement of 'mass' and 'space', which are found:
        a. In the relationship between the solid wall surfaces and the apertures on the façades of buildings;
        b. In the relationship between volumes of buildings and open areas within urban space.
Due to the modification, adaptation and development of urban structures over the centuries, the fourth dimension – time – is now slowly but surely entering the scene.
Today the organizational potential of space has been enriched by two new elements, which will enable us to create a far more comprehensive and more flexible townscape in the future. These elements are:
        a) The mobility and flexibility of man-made structures;
        b) The possibility of achieving total integration of structured and natural space and of creating multi-layered urban formations (subterranean formations plus megastructures).
Between them these two elements will introduce the dimension of time, which will then help to shape the new compositions; the mobility of the structures will ensure constant change within the spatial cluster.

In scenographic frontal formations, in formations with static but modelled forms, and in dynamic arrangements of structured space, the townscape always constitutes a 'spatial' composition on the surface of the earth.

323   *The Townscape as a Sphere of Human Activities*

Apart from its 'visual' and 'spatial' meaning the townscape fulfils a further function as a sphere of human activities. To be more precise, it provides a mirror image of the technological facilities, the aesthetic attitudes and the dominant social structure of every historical period.
If the visual aspect of the townscape represents the conscious attempt to create architectural structures, its social aspect represents the human significance and emotional impact of the urban cluster.

72 and 73

A street in the Plaka, the old town of Athens. Fig. 72 shows the street as it was fifteen years ago. Fig. 73 shows it as it is today. The concrete skeleton of the unfinished building on the left is completely out of place. So too are the wooden masts for the electric street lighting. These incongruous elements have spoiled the townscape.

As a sphere of human activities the townscape both reflects and provides the framework for urban experiences and social activities. This is why so many of the people who live in historic settlements are so attached to their townscape. Nor should this fascination be written off as a superficial and 'romantic' yearning for the past. Its explanation lies rather in the 'discreet message' that is transmitted by historic townscapes and that speaks of the life and urban experiences of past epochs.

As a mirror image of society the townscape naturally reflects the aesthetic attitudes and consequently the architectural style of the period or periods in which it was created. In certain rare cases, we find townscapes which are the product of a single historical epoch, but in the vast majority of cases they are the outcome of the accumulated achievements of successive epochs, which means, of course, that they reflect the historical continuity of human life and of man's cultural development.

33   THE INHERENT PLURALITY OF THE TOWNSCAPE

331   *The Plurality of the Townscape Does Not Constitute a Discordant Factor*

Since the townscape is made up of a great number of different components and since, in virtually all cases[29], it contains architectural styles and spatial conceptions deriving from different epochs, it is only very rarely that it constitutes a homogeneous formation.

However, the juxtaposition of individual buildings, groups of buildings and open areas that have derived from different epochs and were executed in different styles does not necessarily constitute a discordant factor within the total townscape. The restructuring of urban areas, the building of new roads, the creation of open areas and the establishment of new functions within the urban cluster are part and parcel of urban life. Every creative epoch introduces new elements into the traditional townscape, which then make their contribution to the morphological plurality of the urban composition. This is an inherent structural law. *We see therefore that the townscape results from the accumulation of different elements which are not only representative of their own particular period but which also contribute to the total living composition in so far as they correspond to the social and creative needs of later periods.*

---

29   The only exception to the general rule are a few historic urban centres whose morphological homogeneity has been preserved throughout the centuries due either to their geographical situation or to political developments. Examples of such 'timeless' townscapes are to be found in the villages on the islands of the Aegean and the 'medieval cities' in the south of France (e.g. Aigues-Mortes).

Discordant effects are produced, however, if the inherent morphological plurality of the townscape is reduced to a state of visual anarchy by the juxtaposition of incompatible products of different epochs or by the downright desecration of the older urban formations as a result of new urban developments.

The demolition of historic buildings which are beyond repair or which impede new sanitation works (see 431) cannot be regarded as an act of 'desecration'. It must also be admitted that in every creative epoch to date, the emphasis has always been laid on new works of architecture and that scant respect has been paid to traditional creations unless there were special political or religious grounds for preserving them.

Gradual changes wrought in the townscape as a result of the accumulation of interesting architectural entities are also entirely acceptable. But what must be regarded as acts of desecration are the wilful distortion of historic buildings and the impairment of monumental compositions by the introduction of new architectural elements, which are either incongruous or inferior.

74
Another regrettable example of dissonance in a historic townscape: eight-storeyed apartment blocks were erected on the theatre square of Patras (Greece), where they clash violently with the two-storeyed nineteenth-century patrician houses and the elegant fountains.

34   THE IMAGE OF STRUCTURED SPACE IN THE FUTURE

341   *The Revision of the Feeling of Space*

Today we are fast approaching the point where we shall be obliged to effect a revision of our feeling of space. The initial phase of industrialization (1830–1945) has been succeeded by a second phase, in which:

a) New constructional methods involving partial and complete prefabrication have been developed as a result of a systematic analysis of architectural functions;

b) Urban space is being structured in accordance with the findings of an analytical study of urban functions. In this new conception urban space will be composed of 'networks' and 'network centres' which between them will fulfil the various necessary functions of a modern urban cluster (see 221).

In the coming decades two new elements, whose theoretical premises have already been established in some detail, will help to crystallize this new conception of structured space and contribute to its realization. These new elements are:

a) The total industrialization of building;

b) The incorporation into the urban cluster of megastructures and the development of subterranean urban architecture.

75
An example of 'harmonic contrast': two different façades on the Champs Elysées in Paris, which do not clash because they are the same height and have the same ratios between wall areas and apertures. By using the same architectural scale, the French architect has created unity in diversity, which justifies the parallel existence of different architectural forms within the same urban formation.

76
Model of a residential colony constructed from industrialized architectural cells.

77
The installation of the cells in this colony.

78
A six-storeyed spatial grid designed by Yona Friedman as the load-bearing structure for a whole settlement which would have no direct contact with the surface of the earth.

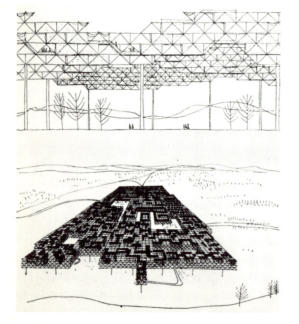

The concept of total industrialization (which should not be confused with prefabrication) can be defined as a new and crucial system of construction based on the use of 'architectural cells'. The industrialized method opens up entirely new dynamic perspectives for the organization of urban space whereas prefabrication merely facilitates and speeds up the work on the building site without changing the essentially static nature of present-day construction.

The architectural cells are made and assembled in the factory from extremely light and durable materials (such as polyester, resin and light alloys) and, when joined together, form honeycomb formations capable of infinite variation which will enable us to change our spatial structures virtually at will.

Voices have been raised of recent years warning us that the total industrialization of architecture must inevitably lead to the demise of creative architecture. But this is simply not true. Far from restricting his creativity, the architect who participates in a multi-disciplinary creative team will find that he has actually increased his effective scope. Of course, this will only be the case if he is prepared to abandon the anachronistic conception of the artist as an 'all-round man' (i.e. the *homo universalis* of the Renaissance) in order to concentrate on his rightful task – the study of functions and of the creation of forms – as a member of a team[30].

---

30 A simple comparison with another sphere of production which is already firmly established is enough to convince us of the truth of this argument. The international motor car industry produces approximately one thousand different models and, despite the fact that these are all mass-produced articles, they nonetheless satisfy a wide range of practical needs and personal tastes. Because of its mass-production methods, this industry is able to make funds available for research into the technical and aesthetic components of motor-car design. Here total industrialization has certainly not undermined the artistic position of the motor-car designer. On the contrary, he is an important part of this whole industrial process. The international reputation of Pininfarina, the Italian motor-car body designer, proves this beyond doubt.

343 Megastructures

With the introduction of megastructures – enormous supporting grid structures which will support whole groups of dwelling units – buildings will no longer have to be in direct contact with the earth and will no longer be subject to the vagaries of the climate.

In the future spatial cluster as envisaged by progressive architects, our residential areas, traffic networks and public squares will be suspended in space by means of gigantic pylons and prestressed cables. *Our buildings will then no longer rest directly on the earth, which will be given over to natural clusters (some landscaped). As a result, the artificial and the natural cluster will at last be completely integrated.*

Large subterranean urban areas will also be developed for communications networks, administrative services and warehouses.

The development of geodesic domes (designed by Buckminster Fuller) and of enormous synthetic and pressurized membranes will make it possible to create artificial micro-climates for whole urban areas.

344 The Spatial Setting[31]

Within the next hundred years, the artificial cluster as we know it today will have been completely restructured thanks to these daring technological achievements. This new kind of structured space will undoubtedly produce a completely new spatial setting.

---

31 'Spatial setting' refers, of course, to the 'structured space' of the future and not to cosmic or interplanetay space.

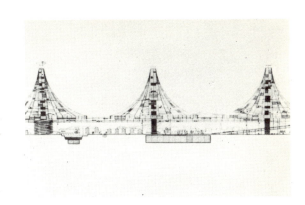

79
Another design for megastructures: Paul Maymont's towers, which make the fullest possible use of space, produce an impression of weightlessness and dispense with the need for direct contact with the surface of the earth.

80
Sketch of a pressurized canopy made of synthetic materials and measuring ten kilometres in diameter. The canopy is designed to cover part of Manhattan and so provide a controlled micro-climate for a whole urban area.

81
Sketch of architectural cells forming an artificial landscape. By using industrial components and erecting them at different levels it is possible to create urban structures which reflect the natural forms of the terrain.

Transition from the Townscape to the Spatial Setting, Principal Characteristics of the Spatial Setting. Dynamic equilibrium of the composition.

The static stability of the traditional townscape with its self-contained, unchanging and unmovable constructions will be replaced by the flexibility, the adaptability, the changeableness and the continuity of the new spatial setting with its open and transformable constructions.

Incidentally, this new spatial setting will dispense with the thousand-year-old antithesis between the urban and spatial clusters, since both will be integrated into the continuum of a new cluster compounded of 'humanized' nature.

345   *The Role of the Historic Centres in the Future Spatial Setting*

Because of the general flexibility of the future spatial setting, two of its components, which will be relatively unchanging, are bound to acquire special significance. These are:

a) The natural landscape which, although subject to seasonal changes, will not change its position within the spatial setting. It will, however, differ from the natural landscape of the present day for, instead of being isolated from the town in the form of a green belt or confined to specific inner areas in the form of parks, it will constitute a 'natural cluster' that will be closely integrated with the human structures[32].

b) The historic urban centres, i.e. all traditional and inhabited urban centres within the future spatial cluster which illustrate the static townscape of the past. Strange though it may seem, the majority of twentieth-century urban formations will feature in the future spatial cluster as 'historic urban centres of recent times'.

The integration of these two static elements into the flexible spatial cluster of the future is not only feasible but also desirable.

It is feasible because it does nothing to hinder the development of the dynamic structures, whose organization presupposes the integration and coexistence of various networks within structured space. It is desirable – and even imperative – because close contact between nature and structured urban space is necessary for man's physical and mental health, while the preservation of restructured historic centres would grant him direct access to his social and cultural heritage and ensure continuity of social development.

We see, therefore, that far from posing a threat to the future spatial setting, the survival of the historic urban centres would actually make a valuable contribution to its richness and diversity. Consequently, their protection, restructuring and rehabilitation are essential tasks, which are a matter of concern primarily for our town and regional planners, but also for those interested in the conservation of architectural monuments.

In view of the urgency of this problem the fourth and final section of this study will be given over to a clarification of the methodology of urban rehabilitation.

32 In this connection see the proposals put forward by Paul Maymont, who envisages metal megastructures 300 metres high to carry residential cells which will thus be suspended above the natural landscape.

4 The Rehabilitation of Historic Centres

4 The Rehabilitation of Historic Centres

41   SPACE AESTHETIC PROBLEMS IN THE PRESERVATION
       OF THE TOWNSCAPE

The preservation of the invaluable townscapes of original historic centres is being ensured by various concrete measures which are being taken by national and local authorities and also by numerous private bodies.

Interventions of this kind presuppose the solution – at a practical deontological level – of certain purely aesthetic problems. On the one hand the historic centres have to be considered as composite urban structures; on the other hand care must be taken to ensure that individual buildings within the protected area are properly treated (i.e. both classified monuments and accompanying buildings of secondary importance).

411   *Aesthetic Problems in Urban Development*

4111   The Urban Structure. The Preservation of the Specific Urban Character
       of Historic Centres

If the historic urban centres are to be regenerated and play their part in the future spatial cluster as centres of private, social and cultural life (see 222), it follows that no development project may be undertaken within a protected area unless care is taken to ensure that the original and unique character of the townscape is preserved.

In nearly all cases, this townscape is characterized by the inherent plurality of its structures. In other words, it embraces a wide range of architectural styles from widely differing epochs which are integrated into a corporate whole. Two particularly striking examples of this kind of formation are provided by Florence and Paris. (In the case of just a few historic centres we find that the townscape stems from a single historical epoch.)

We see, therefore, that the preservation of the essential character of the townscape assumes two main forms, on the one hand calling for the preservation of morphological unity, and on the other for restraint vis-à-vis close-knit morphological diversity. This latter requirement is particularly difficult for it presupposes the parallel preservation

82 and 83
Nineteenth-century monorail in Wuppertal, Germany (82) and a contemporary monorail in Seattle, U.S.A. (83). Great care should be exercised over the introduction of mechanical transport into historic centres. Wherever possible, access should be by way of ring roads and underpasses.

in every part of the total composition of a variety of elements stemming from different historical epochs[33]. The nineteenth century, which was a period of great technological progress and growing urbanization, systematically ignored the specific character of the historic townscape (see 222). The only interest taken in this subject was a strictly limited concern for the conservation of individual architectural monuments, most of which were chosen at random (see 4122).

Meanwhile the urban formations of the historic centres underwent profound changes due to the building of wide new roads, the introduction of tramways and the creation of spacious public squares. Although these installations still contrive to meet the needs of present-day urban communications[34], they should never really have been built, since they were based on the assumption – one that has long since been invalidated – that our historic urban centres would be able to accommodate the communications networks and administrative and commercial centres, not only of our own century, but of the twenty-first century as well. Consequently our historic centres have been subjected to these regrettable and, in many cases, irreversible changes to no real purpose since urban congestion has already reached an absolute peak.

These bitter experiences have taught us that our historic centres must now remain virtually unchanged. In other words, the old traditional urban sectors must be allowed to coexist and interact with the present and future urban cluster (see 433). The greatest possible care should be taken when introducing modern technical infrastructures into historic areas[35], which should in any case be kept to the absolute minimum necessary to provide basic amenities for the population. The same infrastructures must, of course, be seen in present-day urban formations, where they constitute one of the major functional and visual components of the townscape.

84
In the mid-nineteenth century, an extremely daring town-planning project was carried out in Paris when the historic centre was opened up to create new road systems. These were planned by Haussmann at the express wish of Napoleon III. On the plan the black lines represent completed roads and the dotted lines projected roads which were never built.

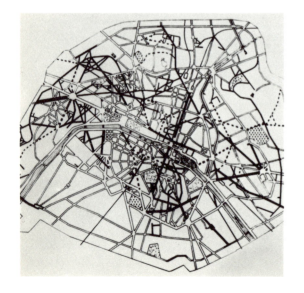

---

33 In the nineteenth century, straight tree-lined roads were built in many historic medieval centres in an attempt to integrate these centres with the new urban districts being erected on their outskirts. This development was a serious error, one that has regrettably been repeated in various European cities.

34 The fact that the streets of Paris and Berlin, which were planned a hundred years ago when the total traffic flow was only a few hundred vehicles an hour, are now able to carry up to ten thousand vehicles an hour would seem to indicate quite remarkable powers of foresight on the part of the planners. In actual fact, however, this fortunate outcome was almost entirely fortuitous since nineteenth century town-planning was largely determined by considerations of pomp and splendour and had little to do with prophetic vision.

35 This question is dealt with in greater detail in Sections 432 and 433.

What then are the principal elements which characterize the townscape of protected areas and which have to be given special consideration in any new development within historic urban clusters?

Apart from the fundamental questions (i.e. whether we are dealing with a 'flat' townscape or with one in which the feeling of space predominates, whether the urban composition is axial and geometrical or free and asymmetrical), the following 'characteristic elements' (which are analysed in 4112–4116) must also be mentioned;

    a) The character of the road network;

    b) The visual density of urbanization;

    c) The integration of different architectural forms within the urban space;

    d) The introduction of various non-architectural elements (signs, advertisements, and illuminations)

into the townscape.

Finally, the problems posed by street lighting and by the use of *son et lumière* within protected areas must also be investigated.

4112   The Characteristic Road Network and its Preservation

Traditional urban space has always consisted of regular rows of architectural structures and 'self-contained' residential buildings with well developed road networks complemented by squares, monumental open areas, esplanades and forecourts[36].

Apart from their practical value as traffic routes and communications systems, these road networks have also fulfilled an aesthetic function, which is no less important. The 'inner' townscape of an urban settlement is composed of a variety of perspective views or, to be more precise, of the successive impressions received by the observer, who moves about on the road network and whose view of the town is therefore largely dependent on the alignment of this network.

The character of every road network is determined by its alignments, which constitute the essential aspect of any road development within the urban cluster.

Both in medieval towns and in small rural centres, which were a product of popular architecture, the roads and paths were invariably extremely narrow. They also followed the natural line of the terrain, even when this involved steep gradients or steps. Further characteristics of these networks were free alignment, constant variations in the width of the roads, small squares and, above all, the total absence of a rectangular grid. Such networks are readily recognizable by virtue of their complete spontaneity.

The Neo-classical networks of nineteenth-century Europe, as represented by Berlin and Vienna (which were the centres of Neo-classicism) and by newly-founded or newly-developed cities such as Washington, Athens and Lisbon, were completely different. Here the broad boulevards which linked the various urban centres were based on a variety of geometrical systems, such as the triangular alignment of Athens, the star-shaped concentrism of Paris and Berlin and the circular organization of Karlsruhe. In these axial compositions we find ourselves confronted with roads of constant width, with gentle gradients and pompous monumentality. In view of this geometrical alignment, we would expect to find that the residential blocks in such cities were set out on rectangular grids. In point of fact, however, this was not the case, for rational design was sacrificed to the desire for ostentation and monumentality which found expression in the 'flat' townscape with its great urban axes and star-shaped squares (see 317).

---

36 Because of the free disposition of buildings in present-day urban compositions (which precludes the possibility of continuous street façades), these well developed road networks are tending to disappear from the modern urban cluster and are likely to be even less prominent in the future spatial cluster.

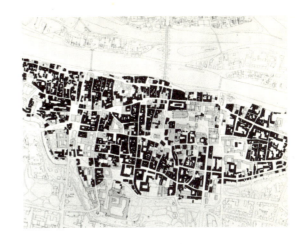

85
Plan of Regensburg on the Danube, Germany. The whole of the historic cluster in this city has been preserved. The winding streets and the freely aligned squares are typical of medieval road networks. Approximate scale 1:25,000.

86
Plan of modern Athens (drawn up by Kleanthis and Schaubert in 1832). The road network in Athens, which follows the geometric principles of Neo-classical design, is quite different from the road network in Regensburg. In Athens we find three principal axes set out in the form of a triangle against a general background of rectangular alignment. Thus the three major roads intersect the side roads at an angle of forty-five degrees. Approximate scale 1:20,000.

87
Medieval street: Great Shambles, York, England. The narrow winding street and the irregular façades are entirely characteristic.

88
Street in the seventeenth-century urban sector of the Marais in Paris with the same characteristics as Great Shambles in York.

Curiously enough, we find examples of completely rectangular alignment (central axis with parallel subsidiary roads and rectangular intersections) in two different historical epochs, far removed from one another in time. The 'new system' employed by Hippodamus of Miletus in antiquity (Miletus, Priene and Selinus) reappeared in the *bastides* of medieval France. Subsequently, this system was employed in some of the ideal cities of the Renaissance (e.g. Freudenstadt) and, finally, in the major cities of the United States (New York, Los Angeles).

It goes without saying that, quite apart from these pure systems, numerous combination systems were evolved over the centuries, usually in major cities. In principle the urban road networks tended to develop either as concentric formations, e.g. in Paris, or as linear formations, e.g. in London.

It was, of course, inevitable that where alterations were made to the alignment of road networks, foreign elements should have been introduced into authentic urban compositions, thus destroying the original 'grid', which provided the general framework for the whole townscape. A striking example of such unfortunate interventions is furnished by the demolition in the mid-nineteenth century of the medieval houses on the Ile de la Cité in Paris in order to enlarge the forecourt of Notre Dame and so ensure that the square answered the ostentatious requirements of the period. Prior to this new development, the enormous height of the cathedral had been emphasized by the urban setting, whose low houses and medieval streets had been dwarfed by its towering structures. Moreover, the small original forecourt had induced a feeling of humility *vis-à-vis*

89

Freudenstadt in the Black Forest, Germany, one of the ideal towns of the Renaissance which was built on a rectangular grid around a central market place. After being completely destroyed in the Second World War, Freudenstadt was reconstructed in accordance with its original geometric plan. The only area that was subjected to arbitrary modifications was the market place, which was rebuilt at various levels to accommodate new traffic routes.

90 and 91

St. Malo in Normandy before the Second World War (90) and after its reconstruction (91). Approximate scale 1:10,000. St. Malo is an example of radical but highly successful restructuring in which a great deal of thought was given to the preservation of the pattern of the historic road network.

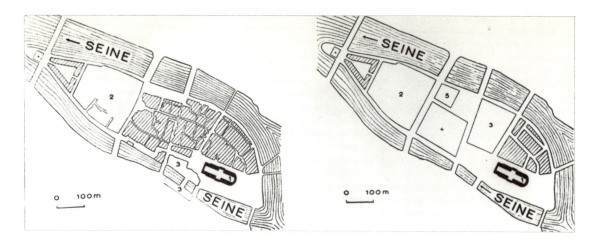

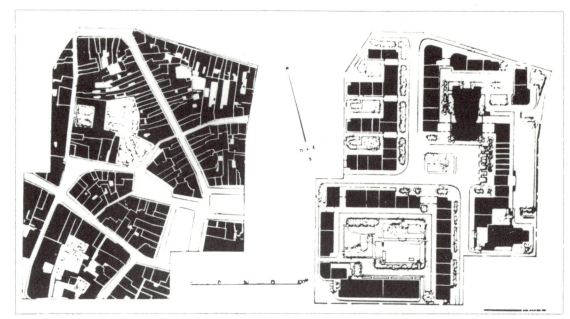

92 and 93
The Ile de la Cité in Paris before (92) and after (93) the alterations to the medieval urban cluster in the nineteenth century. Approximate scale 1:15,000.

94 and 95
Maubeuge, France, before (94) and after (95) its destruction in the Second World War. Maubeuge is an example of bad reconstruction. When the town was rebuilt the original road network was completely altered. Approximate scale 1:3,000.

the lofty Gothic façade which the observer suddenly encountered when emerging from one of the narrow streets that converged on it. Thus the new forecourt destroyed the medieval composition and its psychological effect.

Today, of course, we also encounter misguided attempts to redevelop old towns and medieval urban sectors that are not protected by special regulations. On the grounds of traffic congestion and the need to level out traditional roads, our contemporary planners are intent on laying out the roads and public squares in these old centres in 'regular' patterns.

The consequences of such redevelopment projects have often been extremely regrettable. Freudenstadt is a classical example of unsuccessful and unjustified intervention in a historic town square.

But the alignment of the roads within an urban network is not the only factor that determines the character of an urban cluster. The texture of the materials and the type of road surface also exert a considerable influence.

If the narrow, winding roads of traditional settlements are covered with asphalt, this produces a mental association with the major traffic routes in the heart of the metropolis and gives the impression that these

Aerial photograph of the old district of Gamla Stan, which is situated on a small island in the centre of Stockholm. This historic centre is quite remarkable, for its townscape has remained completely unchanged.

97
Plan of the centre of Stockholm showing the Gamla Stan. Approximate scale 1:20,000.

98
Plan of Beaune, France, showing the fortifications.

99
A characteristic feature of the inner townscape of Beaune are its green areas. In the medieval urban cluster these take the form of compact groups of non-aligned trees.

100
Strasbourg Cathedral with its medieval parvis, which has been preserved in its original form. No trees have been planted on this forecourt with the result that it is still completely authentic.

101
The south façade of Notre Dame in Paris after its modification in the nineteenth century. The gardens, which were added then, completely distort the medieval composition.

102
The Champs Elysées in Paris. Tree-lined boulevards were a typical feature of the townscape of the nineteenth century.

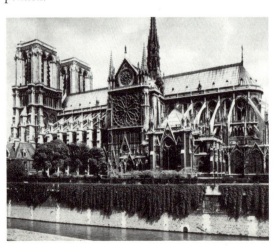

streets are meant to serve the same purpose (motorized traffic routes) as the main thoroughfares, whereas in point of fact they should be reserved for pedestrians (see 434). The best finishes for such roads are cobbles, granite or other stones with similar qualities of durability[37].

Unlike the medieval settlements with their narrow, winding streets, the Baroque and Neo-classical urban centres contained wide open areas, which were situated at the garden sides of palaces and other important buildings. These open areas were either left completely bare or else they were covered with a special sand mix or with fine gravel, which provided a visual transition from the natural sphere of the landscape to the artificial sphere of the buildings, whose inner courtyards were laid out with cobbles or paving stones to emphasize their architectural character (see the square inner courtyard of the Louvre and the atrium of the Zappeion in Athens).

The redevelopment of the forecourt of the old royal palace in Athens, which was carried out in 1930 in order to make room for a monument to the unknown soldier, is typical of the way in which modern planners have changed not only the design but also the texture of monumental open areas. In Friedrich Gärtner's original composition (erected 1836–40), this open area, which occupied a westerly position in front of the main façade of the palace, consisted of a stretch of bare earth with a row of trees at either end. The general view of the palace was greatly enhanced both by the integration into urban space and by the contrast between its architectural quality and the free arrangement of the adjacent open area, which constituted a natural rather than an architectural milieu. By levelling off part of this gently sloping area and erecting a monument to the unknown soldier in front of the enormous retaining wall created by the excavations, the composition was deformed in two different ways. On the one hand the palace lost its visual foundations and, since it now towered up above this enormous six metres high retaining wall, appeared to have been left suspended in mid-air. On the other hand the adjacent open area, which had lost its original form and had, moreover, been covered with stone slabs and walls of porous stone, clashed with the plain and simple lines of the Neo-classical palace.

A further factor which contributes to the definition of the character of urban centres is the planting of trees on pavements. It is widely believed, by experts as well as laymen, that trees on public thoroughfares necessarily contribute to the beauty and originality of a townscape. Unfortunately, this generalization is both naive and false.

The traditional urban clusters of the Middle Ages contain very few trees. Small groups of extremely ancient trees are found on medieval squares, where they constitute what might be described as natural *objets trouvés*, which act as focal points of interest and so contribute to the formation of the townscape. But, apart from these isolated groups, there are virtually no trees in medieval centres for the simple reason that the narrow streets are unable to accommodate them. In the case of Notre Dame in Paris, for example, it is perfectly obvious that the romantic gardens on the south side of the cathedral and of the new forecourt are a late addition, for they clash violently with the medieval composition.

In the urban formations of the Renaissance, Baroque and Neo-classical periods, on the other hand, public gardens were treated as an organizational and architectural component of town planning and were fully integrated with the buildings. Consequently, the elimination of such gardens in order to create public squares or to improve the traffic flow greatly undermines the unity of the whole urban formation.

The rows of trees (plane trees and horse chestnuts), which lined the great boulevards of the nineteenth-century urban sectors, also formed an integral part of the urban composition of that period and, as such, made a major contribution to the 'flatness' of its ostentatious townscape. It is even arguable that the contribution made by these endless rows of trees, which were invariably included in the original plans for any new boulevard, was as great as that made by the monotonous terraced façades.

103
The main façade of the old Royal Palace in Athens after the open area in front of it had been converted into Constitution Square in 1930.

---

37 A few years ago it was foolishly decided that the streets of the Plaka – the old part of Athens where asphalt roads were far from customary – must be 'improved', i.e. asphalted.

The old Royal Palace in Athens with the gently sloping open area in front of its main façade as designed by Friedrich Gärtner. Photograph taken in 1870.

105
Recent aerial photograph of the old Royal Palace in Athens showing the new square with the monument to the unknown soldier (1930).

In conclusion, it bears repeating that any town-planning developments involving alterations to the alignment and texture of historic urban road networks or the planting of trees on those networks should only be undertaken after mature deliberation and should always be kept to a minimum. All such operations call for a thorough knowledge of the particular townscape under review if its essential character is not to be distorted by incongruous additions. It is only in this way that the morphological continuity of protected areas can be ensured and the spirit of their original urban composition preserved.

4113   Visual Density of Urbanization. Exploitation and Settling of Land

The second characteristic element of the historic urban centre might be described as the 'method of urbanization', i.e. the disposition of individual buildings and their reciprocal relations with open areas (public squares, inner and outer courtyards) and with the road network.

The observations made in this section apply to the great majority of historic urban centres (medieval towns, Renaissance towns, and rural settlements containing examples of popular architecture) and should enable us to grasp the essential characteristics of their townscapes. There is one group of historic centres which is not covered by these remarks, namely the urban formations which were developed in the eighteenth and nineteenth centuries in the heart of the major European cities. These were the first urban areas to be subject to 'planning regulations', which stipulated a regular alignment of buildings of more or less equal prominence.

Town planning – which only came into existence as such at the end of the nineteenth century and whose importance increased in direct proportion to the growth of urbanization and the development of the great European cities[38] – evolved basic concepts which have now been embodied in various planning regulations and which impose a somewhat rigid framework on all modern urban formations. *These basic concepts are the density of the population in any given suburb[39], the total useful capacity of buildings[40], the architectural density of a given site[41] and the maximum height of buildings[42].*

By establishing an architectural scale, by controlling the use of land and by taking measures which ensure satisfactory living conditions – regulation of density, creation of open areas, provision of adequate light and sun in residential districts, and stipulations of minimum distances between individual buildings – the town-planners have provided the necessary framework for the settling of urban areas. Because of this legal 'framework', twentieth-century towns and suburbs undoubtedly enjoy great technical and social advantages, especially in the sphere of hygiene. On the other hand they have also suffered considerable losses, for by comparison with the aesthetic diversity of our historic settlements, the new urban areas seem sad, dreary places, due largely to the lack of plasticity in their general design.

---

38 The fundamental tenets of scientific town planning were first formulated by Camillo Sitte in his *City Planning according to Artistic Principles* (Vienna, 1889).

39 The total number of inhabitants divided by the total area (built-up area, road network and public squares) measured in hectares. As a general rule this figure falls between fifty and four hundred inhabitants per hectare, although in certain overcrowded sectors of major cities it reaches one thousand inhabitants per hectare.

40 The total useful capacity of any given parcel is the total useful area of all storeys of the building erected on the site, divided by its total area. This figure varies from 10% for the buildings in certain central districts to 0.2% for the residential buildings of outer districts.

41 The architectural density of a given parcel is the built-up area of the site measured as a percentage of its total area. This figure varies from 100% for central, commercial and shopping areas to 15% for outer garden suburbs.

42 The maximum height of buildings in urban areas (which is measured from the mean level of the base of the principal façade to the highest point of the roof) is fixed by the planning authorities. It is measured in metres and varies considerably.

106 to 110
Examples of 'inner' and 'outer' townscapes which illustrate the visual density of urbanization found in many historic centres.

106   Narrow street in the old town of Chios in Greece.

107   Narrow street in Pylos in the Peloponnese.

We are forced to the regrettable conclusion, therefore, that over the past one hundred years the originality, freedom and creative spontaneity of urban design – which is still found in the urban clusters of historic settlements – has been sacrificed for the sake of technological progress and improved standards of hygiene. The pity of it is that this indisputable advance in the organization of social life need not have been accompanied by such unimaginative and monotonous townscapes. None the less, there have been a number of encouraging developments during the last two decades, which would seem to indicate that technological organization and creative fantasy may well be combined in the near future.

Because the population density of historic towns and districts is subject to such wide variation – due partly to the great diversity of the functions fulfilled by the individual buildings and partly to the fluctuation in the living standards of individual families – no attempt has been made to impose an upper limit. Consequently, population density is not a characteristic element of historic centres. *But the visual density of the buildings which go to make up a historic centre certainly is a characteristic element, which we propose to refer to as the 'visual density of urbanization'.*

The free and 'continuous' disposition of the buildings and groups of buildings in historic centres, which is characterized by the irregular line of the street façades, the free alignment and the arbitrary layout of the road network, the repetitive forms of the rooftops and terraces which, although almost equal in size, are built at different levels, the ever-changing interplay of buildings and open areas (courtyards and gardens), invariably create an impression of density and of structural and formal cohesion. *This impression of cohesion made by the total composition is what is meant by the 'visual density of urbanization'. It is an important element and one that must be protected when new developments are undertaken within historic settlements.*

Any free parcels or building sites created by the demolition of derelict buildings that are situated either on the outskirts of historic centres or in the immediate environs of medieval fortresses can be used for small squares, open areas or belvederes. But if such sites are situated inside the urban cluster, then it would be preferable to use them to erect new buildings (constructed in accordance with the principles of 'harmonic integration' or 'harmonic contrast' – see 4123) so as to preserve the architectural scale and the 'visual density of urbanization' of the original historic townscape.

If we consider the 'total useful capacity', the 'architectural density' and the 'maximum height' of the buildings in historic centres we find that, with the exception of the districts built in the eighteenth and nineteenth centuries, no attempt was made to exploit the available sites within the historic centres in a uniform and rational way. In the complete absence of planning regulations, private property owners were able to do as they pleased and the settlement and exploitation of land, which were pursued in a completely arbitrary manner, often produced extreme contrasts[43].

Although the methods of settlement and land exploitation pursued within historic centres varied enormously, this did not undermine the cohesion of the urban cluster. On the contrary, it strengthened the impression of unity and ensured 'visual density of urbanization'. Here too the aesthetic principle of 'multiplicity in unity' was demonstrated to the full.

43 For example: on one building site a two-storeyed house was built with its principal façade overlooking the street and later single storey workshops were erected immediately behind it. Meanwhile, on the neighbouring site, a three-or four-storeyed house was built with its façade set back from the road to allow for a small front garden. The ground to the rear of this building was then given over to a courtyard and garden.

Although they had no planning regulations to guide them, the builders and architects of past centuries had a highly developed aesthetic sense and voluntarily observed the unwritten regulations of traditional practice. For example, no building was erected in the immediate vicinity of important religious or secular monuments for fear of reducing their visual impact. Moreover, the perspective views of the most important points in the urban and natural cluster were always zealously protected and no new developments were ever undertaken which were likely to impair the stimulating urban experiences which these views provided.

The freedom and diversity of the structural organization of our historic centres are amongst the most valuable features of urban design and must be preserved. Our protective legislation should be directed to this end.

For the past one hundred and fifty years our architects and town-planners have been obsessed with the idea that buildings must be organized to produce regular 'street façades' and that the settlement of urban areas and the exploitation of land must be based on a uniform design. In the new developments which will have to be undertaken in our historic centres, it is imperative that this outdated concept be abandoned.

108 Tossa del Mar on the Costa Brava in Spain.

109 Saint-Tropez in the French Riviera.

110 Bonifacio in Corsica.

111
General view of the Plaka, the eighteenth- and nineteenth-century town of Athens, on the northern slopes
of the Acropolis. Since there were no planning regulations in force when this town was built, it developed in
accordance with the unwritten, i.e. traditional, laws of
urban composition. One of its principal features is the
variation in density between different sites.

4114   Archaeological Sites and Historic Urban Centres. Questions of Formal Diversity in Urban Space

The characteristic accumulation of architectural forms and structures, which was gradually transformed
in the course of the centuries, is also found at a completely different level in urban compositions, where
it helps establish the formal diversity of the historic townscape. This formal diversity – which stems from
the coexistence and confrontation within the historic urban nuclei of groups of buildings from different
periods – need not be a discordant factor. Present-day interventions in protected areas should try to
give all the different architectural styles their due, since this is the only way of preserving the traditional
townscape.
If we wish to preserve the clarity and unity of a particular architectural style by restoring certain buildings
to their original form, we find that, in certain instances, the buildings have to be 'stripped'. But, although
this procedure is justifiable in the case of individual buildings, it cannot be applied to a whole historic sector, whose morphological plurality is not only inherent and unavoidable but also desirable.

112 to 115
Examples of the visual density of urbanization in historic centres.

112 Detail of Nauplia in Greece. Photograph taken from the fortress.

113 Houses on the quayside of the historic harbour of Honfleur in France.

114 The cathedral and historic centre of Strasbourg.

115 Reconstructed model of the medieval town of Nürnberg in Germany.

However, the plurality of the historic townscape does pose a special problem which constitutes a threat not only to the cohesion but to the very existence of the traditional urban cluster. The major difficulty here is presented by the need to integrate individual archaeological monuments or archaeological sites into historic urban centres, for this frequently involves an extremely difficult decision as to whether the protection of an inhabited historic settlement should take priority over archaeological excavations or vice versa. If the decision goes in favour of the archaeologists, the survival of the settlement may well be called into question.

116

Old engraving of the town of Arles showing the Roman amphitheatre, which shows medieval buildings which have been erected on it. This engraving clearly demonstrates the close integration of buildings from different epochs in the medieval townscape.

117

The centre of Arles today: the amphitheatre has been cleared of its medieval buildings and, like the Roman theatre, has been integrated into the dense urban cluster of the medieval town. These two Roman remains have also been given new functions: the amphitheatre is used for bullfighting and the theatre for open air performances in the summer months.

118

The Roman mausoleum of Emperor Galerius (known as the Rotunda) in Salonica has now been converted into a Greek Orthodox church and dedicated to St. George. Previously, the mausoleum had served as a mosque, hence the minaret. Despite these changes of function, this Roman building has retained its symbolic significance and still serves as a point of orientation within the urban cluster.

119

The Hephaesteion (Temple of Vulcan) in Athens and the site of the classical agora prior to 1930. At that time the agora was still covered by houses.

120

The same site after it had been excavated (1930–60) and converted into an archaeological park in 1960. In this particular case the decision to sacrifice part of the old town in order to carry out archaeological excavations of a classical site of major importance was fully justified.

121

The historic centre of Rome. In the foreground, the monstrous monument to Victor Emmanuel, which disrupts the perspective view of the Forum Romanum from the Piazza Venezia. On the background the Forum Romanum with the triumphal arches erected to Titus and Septimus Severus and the Colosseum. The antique centre of Rome was never covered by new urban sectors. This we know from the engravings and town plans of the Renaissance.

Individual archaeological monuments are often well preserved and can still fulfil specific functions within the urban cluster. There are many examples of successful integration in this sphere. The Pantheon in Rome was transformed into a church, the Diocletian baths now serve as a museum, the Roman amphitheatres in Arles and Nîmes are used for open air performances and until one hundred and fifty years ago Greek Orthodox services were held in the Hephaesteion (Temple of Vulcan) in Athens. Other archaeological monuments, which have survived as ruins, have acquired an aura of dignity as witnesses to the past and, because they are so intimately connected with their neighbourhood, these uninhabited and functionally 'dead' monuments constitute an integral part of the townscape and so fulfil an important function as symbols of urban continuity.

We see, therefore, that the integration of an individual archaeological monument that is situated in a historic milieu poses no special problems provided its present functions, or its symbolic significance, or both of these things together, enable it to play an essential role as part of the townscape and of the urban composition.

But archaeological sites are a very different matter, for they pose deontological problems which are often insuperable. For example, we find ourselves faced with a painful alternative when living towns of inestimable historic value are found to have been built on the site of important archaeological ruins. Someone has to decide whether the living historic centre should be preserved and the archaeological ruins left untouched, or whether excavations should be carried out and part or all of the historic settlement destroyed. It is extremely difficult to determine which of these two is the more important in cultural terms, even when the advice of artists and scientists of absolute integrity is available. Subjective preferences are bound to play their part. Although each case must be judged on its merits, the ultimate decision will depend to some extent at least on the following factors:

a) An objective comparison between the living historic centre and the archaeological site to try and establish their aesthetic and cultural values.

b) An assessment of the uniqueness of each of these two compositions. This would involve an enquiry to discover whether there were other historic centres of the same period and with the same stylistic qualities as the centre in question, and a similar enquiry to discover whether there were other archaeological ruins essentially the same as those which it was hoped to excavate.

c) A positive expectancy – based on documentary evidence – that the proposed archaeological excavations will yield significant finds.

d) The symbolic and cultural repercussions of the survival or destruction of the historic centre on both its urban and its rural environment.

For the experts or the general public to exaggerate the importance of either of these two complexes (the 'living' historic and the archaeological complex) would be a serious mistake, for this would make for an arbitrary choice and constitute a denial of scientific deontology.

The respect for and systematic study of Prehistoric, Classical, Hellenistic and Roman antiquities, which was characteristic of early nineteenth-century Greece, contributed to the international interest in this field which was sparked off by the important archaeological excavations carried out from 1860 onwards. Unfortunately this interest soon became exaggerated and led to a sort of 'archaeological mania'. As a result, the living architectural tradition of the Greek nation – which still plays a leading role in the spheres of music, craftsmanship, painting and folk poetry – was completely ignored, which meant that the Greek architectural monuments and historic centres of the last five hundred years were grossly underestimated. It is hardly surprising, therefore, that in Greece the decision to sacrifice living historic settlements to archaeological excavations has been taken far too easily.

In the Germanic territories of Western Europe, on the other hand, where antique remains are relatively rare (they consist primarily of basilicas, baths, fortresses and a small number of Roman theatres), it was a long time before the idea of excavating architectural ruins in the nucleus of a historic settlement was accepted. One reason for this attitude was doubtless the latent aversion felt for the remains of a foreign culture. But the principal reason was almost certainly the greater sensitivity demonstrated by the peoples of Western Europe *vis-à-vis* their traditional townscapes, which might have suffered if archaeological excavations had been carried out in their urban centres.

We must now concern ourselves with another important aspect of this subject, namely the function of existing and projected archaeological sites within the sphere of historic towns and the integration of these very special open areas into the urban cluster.

In this particular case the need to protect the scale and unity of the townscape must take precedence over all other considerations.

If it is proposed to carry out archaeological excavations in a living historic urban sector that is situated on the outskirts of an 'old town' this should be allowed, for although the townscape would be changed, neither the cohesion nor the continuity of the urban cluster would be disrupted. After the excavations have been completed, the site could be converted into an archaeological park, where the excavated ruins could be put on show[44]. The park and the historic centre would then be able to coexist without detriment to one another.

But if it is proposed to carry out deep excavations of sites situated in the heart of a historic centre, this would certainly threaten the cohesion of the townscape. Large historic urban centres, which contain imposing groups of buildings, are of course better able to sustain this threat. (The Renaissance and the Baroque city of Rome, where antique remains such as the Colosseum and the Forum have always formed an integral part of the townscape, is a case in point.) The living urban cluster suffers less when important antique monuments are left standing on their original sites, where they form a complement to the historic centre. The presence of trees and a hilly terrain also facilitates the integration of archaeological sites into historic centres.

But when excavations are carried out in a relatively small historic centre which has been built on fairly flat ground, then the cohesion of the urban cluster is far more likely to be disrupted. The Roman agora and the Library of Hadrian, which are situated in the heart of the Plaka, the old town of Athens, provide a striking example of this danger. Up to one hundred and fifty years ago – i.e. before the excavations were started – the Tower of the Winds (Cyrrhestes' hydraulic clock) was the only antique building in the old town which formed an integral part of the urban setting. But today the exploratory excavations which have been carried out in the vicinity of the Tower completely disrupt the unity of the historic townscape. A relatively narrow area has been dug, with the result that the archaeological ruins, which have been only partially excavated, present a somewhat sorry sight. The only possible solution would be to extend the excavations and expose the whole of the Roman agora. In this way the Roman and classical agoras could be integrated with one another to form a composite site on the outskirts of the old town.

It follows, therefore, that individual archaeological monuments and sites can constitute a positive element in the townscape of the historic centre. This is not the case, however, when large and monotonous sites are created, since these must necessarily disrupt or even destroy the cohesion of the urban cluster.

122 and 124
Detail (122) and plan (124) of the archaeological site in front of the Odeon of Herodes Atticus in Athens, which has been converted into an archaeological park.

---

44 The conversion of archaeological sites into archaeological parks is a very difficult undertaking and calls for great powers of empathy on the part of the landscape architect. Trees and shrubs have to be kept under control to prevent the archaeological ruins from being engulfed by the vegetation while paths, steps, retaining walls and any other elements used by the architect for his landscaping must be clearly distinguishable, in both form and substance, from the archaeological remains. Three interesting examples of archaeological landscaping are to be found in Athens. They are:

a) The approaches to the Acropolis and its neighbouring hills (Pnyx and Philopappos). Architect: Demetrius Pikionis; work carried out between 1955 and 1958.

b) The conversion of the archaeological site in front of the Odeon of Herodes Atticus. Architects: Alexander Papageorgiou and Christos Lembessis; work carried out in 1959.

c) The conversion of the archaeological site to the north of the Olympeum. Architect: John Travlos; work carried out in 1962.

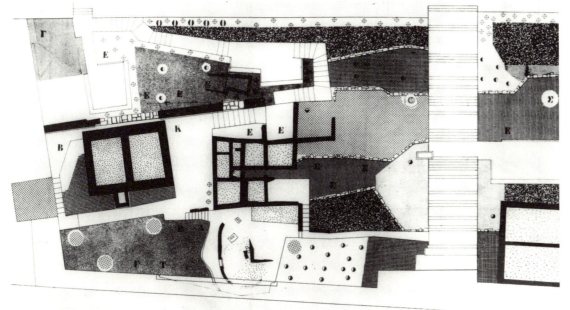

123
Panoramic view of the archaeological site of the classical agora in Athens from the south. From left to right: the Hephesteion, the agora, the Stoa of Attalus and part of the Plaka.

125
Panoramic view of the same site from the west with the classical agora in the foreground and the Plaka in the middle ground. This picture shows the two link mechanisms which ensure the visual transition from the modern city centre of Athens (left) to the antique monuments on the Acropolis (right). They are the archaeological site and the Plaka. With its modest scale and its central position on the slopes of the Acropolis, the Plaka is ideally suited to this purpose.

126 to 129
The partially excavated Roman agora in Athens. This is an example of an archaeological site within a historic urban cluster.

126   View from the west. On the left the mosque, on the right the 'Tower of the Winds'.

127   View from the east. In the background the western entrance to the agora. On the right the nineteenth-century burghers' houses.

128   Part of a plan of the Plaka showing the classification of different buildings:
Black areas with white dots: contemporary buildings.
Black areas: buildings of great architectural value.
Dark grey areas: accompanying buildings.
Pale grey areas: inner courtyards and gardens.

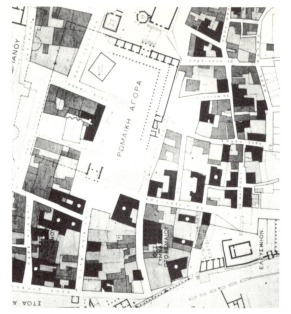

129   Another view from the west. The Ionic columns of the Roman agora are clearly visible.

130 to 133
Example of archaeological sites on the outskirts of a historic centre.

130   The classical agora in Athens on the edge of the Plaka.

131   The Olympeum and its environs on the eastern edge of the Plaka.

132 Plan of the Acropolis, the Plaka and the archaeological sites. The classical agora is on the left, the Roman agora and the Library of Hadrian are in the top centre. Approximate scale 1:6,000.

133 Aerial photograph of the same area. Whereas the classical agora, the Temple of Zeus and the monuments on the Acropolis do not disrupt the unity of the townscape of the Plaka in any way, the Roman agora and the Library of Hadrian have left undesirable gaps in the centre of this urban cluster. Approximate scale 1:6,000.

134 to 136
Artefacts which are more often encountered in museums than in urban clusters.

134   Old battleship in the historic harbour of Honfleur in Normandy.

135   Pre–1914 Benz motor car.

4115   'Local Colour' and the Falsifications of Urban Displays.
       Advertising, Signs and Lighting in Protected Areas

The quality of the townscape of historic centres is also influenced to a considerable degree by the decorative forms and elements that are found on the streets and on the façades of buildings in protected areas and by the quality of the communal 'objects'. These may be in keeping with the townscape or they may clash with it.

If we accept that the protection and regeneration of historic centres is a worthwhile undertaking, we still have to ask ourselves whether every single object found in a protected area (i.e. the type of lettering used on signs, the various kinds of traffic signals, and the traditional styles of lamp standards) should remain unchanged.
In our view such an extreme attitude would be quite utopian. If, for example, it was decided that the only street lights to be allowed in a certain area were to be the sort of gas lanterns used in 1850, then it would logically follow that all vehicular traffic would have to be horse-drawn and that all the inhabitants would have to wear mid-nineteenth century costume so as to preserve the 'atmosphere' of their historic centre.

Such ridiculous conservative ideas would soon get us into serious difficulties. The reason why historic buildings have survived is that they have continued to fulfil their original function. Their present form, even when it is more or less homogeneous, was evolved over a long period of time and embraces stylistic and material elements dating from different periods. And so, if we were to try to apply this absurd principle of preserving in their original form the entire range of 'objects' which are to be found in the historic townscape, we would first have to decide which historical style was to be the dominant style and then make every single object in the urban cluster conform to it. Should we choose the early twentieth century (veteran motor-cars, gaslighting and Art Nouveau) or the early nineteenth century (horse-drawn carriages, candelabras and Neo-classical signs)? Should we choose some other period or should we opt for a scenographic combination of several periods? The problem is, of course, insoluble and it is not difficult to envisage the sort of absurd falsifications to which such arbitrary decisions would give rise.
Even if such an eclectical townscape were created by an architect of talent, it would still be an artificial structure that would never constitute a living townscape and would, therefore, have no relevance for either the present or the future. We have already drawn attention to the 'timelessness' of historic buildings, i.e. to their ability to go on fulfilling functions which answer specific and essential human needs. This is not to deny the importance of the revolutionary building techniques which have evolved primarily in the course of the past fifty years, or of the new industrial processes which will soon place at our disposal entirely new structures or 'containers of human functions' that will diverge more and more from our traditional constructions. But these traditional constructions can nevertheless be rehabilitated since they are ideally suited to man's residential and cultural needs, which have hardly changed throughout the centuries and are likely to remain essentially the same in the future.
It is often maintained that since the preservation and systematic use of old objects in historic centres constitutes a falsification, the preservation of the actual centres must be a romantic and illusory undertaking[45]. This argument is, of course, based on a fallacy, for there is a fundamental and essential difference between the objects and the buildings in a historic centre.
Although the systematic retention and use of traditional objects in historic urban centres as a means of establishing a specific historical style is completely utopian, this does not mean to say that traditional objects should not be used at all. On the contrary, such practices should be encouraged, if there is a functional meaning to them.

---

45 See section 222 for a detailed discussion of the role and the site of the rehabilitated historic centres in the future spatial cluster.

Provided the gas mains are still intact there is no reason why gas should not be used for street lighting, in which case the traditional and, in many cases, stylistically extremely interesting gas lamps might also be retained. (Whole areas of Berlin are still very effectively lit by gas lamps.) The townscape of historic centres would also be enriched by the presence of horse-drawn vehicles and veteran and vintage motor cars still in existence, provided these were not specially introduced as 'showpieces'. Old commercial and shop signs should certainly be retained.

But our historic centres are not only concerned with the past. They also have to incorporate modern functional objects, and the important thing here is to ensure smooth and harmonious integration.

The question of motorized traffic in historic centres will be analysed in detail in the further course of this study (see 433 and 434). Consequently, we now propose to deal only with the difficult problem posed by the development of new methods of illumination and new types of signs.

The way in which new objects are introduced will depend on which of the three methods governing new creations in protected areas (see 4112) is adopted. These methods involve:

a) Exact copies[46];
b) New creations which aim at harmonic integration;
c) New creations which aim at harmonic contrast.

---

46 These copies should not be regarded as falsifications. They are artistic copies of movable objects within historic centres and, as such, constitute a legitimate method of duplication.

136 Early nineteenth-century lift in a boarding-house in Berlin.

138 Paris. In front of the Louvre.

137 Paris. Place des Pyramides.

139 Berlin. A street in the city centre.

137 to 139
Authentic gaslights which have been converted to electricity and have consequently been preserved.

140 Athens. Acropolis district. Street sign that is completely out of place in this historic area.

141 Berlin. Shop window.

142 Athens. Neo-classical patrician houses whose façades have been all but blotted out by signs.

140 to 145
Buildings and streets defaced by the arbitrary installation of unaesthetic signs.

143 Honfleur in Normandy. Advertising sign on an attractive slate façade.

144 House façade in Pylos in the western Peloponnese.

145 Paris. Shops on the Champs Elysées.

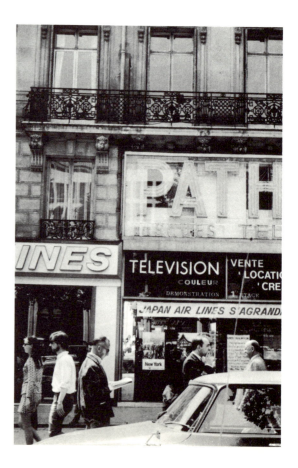

As far as signs are concerned, the type of lettering employed is unimportant; it can be taken from the traditional repertoire or it can be a piece of extremely modern graphic design. However, the size, the colour and the layout of the sign do matter; these should be unobtrusive and in keeping with the façade of the historic building on which the sign is to appear. The glaring colours, arbitrary shapes and large formats, which are so popular with our more energetic and enterprising business men, should be rejected. The positioning of these commercial and shop signs must also be carefully thought out so as to fit in with the architecture of the façades, thus ensuring that the buildings are not deformed. (Friezes and the wall-panels between doors and windows would be suitable.)

The illumination of these signs poses an additional problem since it too constitutes a possible threat to the townscape. Unfortunately we often find that unsuitable forms of lighting are tolerated, not only by the public, but even by the authorities. Flashing neon signs should not be allowed in historic centres. They constitute a disturbing and undesirable element in these peaceful areas of urban life.

Moreover, the use of such neon lights would make it impossible to create artistic lighting effects for important houses and monuments. Consequently, the authorities should insist on indirect lighting for all signs in protected areas.

146 to 148
Cases where the townscape has been respected by the installation of suitable signs.

146 Old town of Zürich. Sign correctly fixed to a frieze.

147 Old town of Zürich. The signs and flowers on this building actually increase the aesthetic appeal of the façade.

148 Gamla Stan, the old town of Stockholm. A public convenience which constitutes an integral and decorative part of the townscape.

149 The market place of the small town of Orchomenos in Boetia, Greece. This small square contains works from three different periods: a small Byzantine church, a provincial burgher's house dating from the nineteenth century and a wrought iron belfry executed in a twentieth-century style. The 'modern' lamp standard is the only disruptive element.

150
Pylos near Navarino in the Peloponnese. The townscape of this historic settlement has been defaced by lamp standards and electric cables.

Indirect lighting should also be used for the illumination of streets in historic sectors. The one exception to this general practice would be in the case of the great nineteenth-century boulevards and intersections, where modern street lamps actually have a beneficial effect, provided they are sited between the trees which normally line these routes. The erection of old-type gas or oil lamps fitted with electric bulbs is patently absurd, although it may perhaps be condoned if the form of the lamps is authentic. All too often such lamps reveal a very inadequate sense of form. As for the practice of mounting electric lights on masts or cables, this sets up a positively barbaric and highly discordant contrast to the historic townscape.

For the indirect lighting technique advocated above, floodlights of varying strengths would be mounted in windbreaks, niches and projecting cornices on the façades of buildings where they would not be seen by passers-by. These floodlights would be trained on the buildings and would, therefore, also provide indirect lighting for the streets below.

Apart from commercial signs and street lamps, other private and public objects which influence the streetscape would have to be kept under surveillance to ensure that the appearance of the historic townscape was not spoilt by the introduction of inferior elements. These objects include street signs and house numbers, verandas, benches, decorative railings, cellar entrances, kiosks, shelters and floral displays.

The specific townscape of historic centres, whose characteristic elements have been analysed above (see 4111 to 4115), is often thought of by the general public in terms of 'local colour'. This concept has given rise to serious misunderstandings. The regrettable and completely irresponsible desire for a quick profit, which is rendered even more pernicious by a total lack of culture, has frequently prompted business-men, restaurateurs and even private individuals to 'upvalue' their traditional urban image by conjuring up a special kind of 'local colour' based on their own personal taste and ideas. This unaesthetic and lamentable tendency has found its principal outlet in a sort of 'pseudo-architecture', in which flimsy materials (fibre board, plywood and lathes) are used to produce what are virtually theatrical decorations. By turning out bad copies of traditional styles and painting them in glaring colours, these would-be designers think that they can 'embellish' their historic milieu. Some are quite prepared to transform whole historic buildings in this way[47].

The creation of 'local colour' has posed a greater threat to the historic centres than either the ravages of war or the demolition squads of the land speculators. Their assault was direct and consistent and it led to the complete elimination of specific historic centres. The effects of the 'local colour' movement are more insidious and threaten to degrade many such centres.

---

47 This tendency is constantly encountered in the historic sectors of big cities, which attract large numbers of tourists. The major problem in the formulation of special building regulations for such historic sectors is how to prevent the spread of this calamitous and completely mistaken conception of 'local colour'. The lamentable deformation of the Plaka in the course of the past ten years and the partial deformation of the old part of Zürich (which is otherwise completely intact) afford striking examples of the harm done by unfettered private initiative. By contrast, the historic urban sectors of St. Germain and Montmartre in Paris have been successfully protected right down to our own day.

In the course of the past twenty years important progress has been made in respect of the illumination of individual architectural monuments and other architectural complexes. Various big European manufacturers of electrical equipment (Siemens, Osram, Philips and AEG) formed research teams which have developed revolutionary and highly artistic techniques for use in this sphere.

During the first phase of this research work, the method of illumination employed was extremely powerful but quite static and undifferentiated. Architectural monuments of great symbolic significance were simply exposed to a sea of light, which fell on them from all sides and was supplied by powerful floodlights set out in an arc around the building. The relief modelling, the details of the façade and all the other architectural refinements were sacrificed to this rudimentary form of illumination. What is more, the adjoining urban sector was greatly distorted by these blinding floodlights[48].

Static illumination was still used during the second phase of this research, albeit with far greater care. The harsh illumination in general use until then was replaced by a differentiated system, in which several kinds of floodlight were trained both on the monument itself and on the buildings in its immediate vicinity. These floodlights, which varied in strength and were arranged in groups of various sizes, also varied in respect of colour. The full range covered all shades of white from the cold, hard white of neon lighting to the dark yellow of sodium. This well considered technique, which obtains its effect by subtle nuances of light and colour, is able to meet the requirements of all kinds of monuments.

Apart from this static technique a new, dynamic system of illumination has been evolved during the past ten years, which is calculated to enhance the aesthetic effect of architectural monuments. This system, which was developed primarily in France, has become known as *Son et Lumière*. It has been used chiefly for the illumination of important groups of buildings and has frequently helped to bring out their monumental quality, although on certain regrettable occasions it has quite the reverse effect and actually leads to the desecration of monuments.

'Son et Lumière', in which the setting is provided by an authentic architectural monument, lies midway between an actual scenic performance and a display of lighting effects. The originality of this technique lies in the fact that there are no actors, their function being taken over by the suggestive and ever-changing lighting effects, which bring the monument to life, and by the background music and historical commentary, which provide a backing for the illuminations.

Son et Lumière performances have been staged in a wide variety of historic settings including the Pyramids of Giza, the Acropolis in Athens and the Palace of Versailles.

Such performances serve a dual purpose: they enhance the aesthetic quality of the historic building and they recall the historic events which have taken place in it. Not surprisingly, therefore, the technique is best suited to medieval and modern settings, whose history is still comparatively well known and whose nocturnal atmosphere is more easily conjured up by means of illuminations.

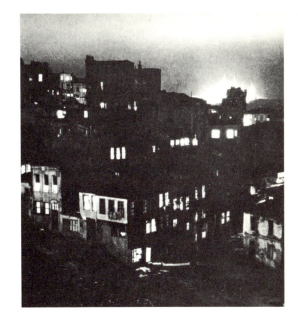

151
Constantinople. The old town with the Suleiman mosque in the background lit up by powerful floodlights. This 'static' and undifferentiated type of floodlighting is not often used today.

152
The floodlit Suleiman mosque. Another view.

---

48 A typical example of this kind of illumination was that provided for the Parthenon in Athens. The illumination employed up to 1960 was so powerful that the temple appeared to have lost all contact with the earth and looked more like a gigantic floating doll's house than a monument crowning a city.

153
One of the phases of the 'Son et Lumière' performance on the Acropolis in Athens.

154
Static illumination of the Cathedral of Barcelona. By the use of a large number of floodlights, delicate effects have been obtained.

155 and 156
Two phases of a 'Son et Lumière' performance at the Cathedral of Lisieux in France.

4121   General Remarks. Classification of Individual Buildings

We have already established that the fundamental criterion to be applied in the 'rehabilitation' of historic settlements is the original creative idea underlying its composition. Even so, we are still faced with a wealth of aesthetic and practical problems which have to be solved if we are to establish specific criteria for the treatment of buildings (classified architectural monuments and other structures) in protected areas.
The 'treatment' of individual buildings for purposes of protection raises serious deontological problems which are aggravated by the fact that, far from being isolated, historic buildings almost invariably form part of a larger group. Consequently, they contribute, both as individual structures and by virtue of their interaction with neighbouring buildings, to the characteristic 'townscape' of their historic urban area.
The buildings in historic settlements fall into three principal categories:

a) Architectural monuments which have either already been classified or have been accepted for classification.

b) Accompanying buildings of limited aesthetic value and limited historical and symbolic significance. These have usually survived in relatively large numbers and are usually of the same basic type (e.g. the burghers' houses of the fifteenth to seventeenth centuries). But, although they cannot be regarded as monuments, such buildings interact with the more valuable historic monuments to form a morphological entity that is a constituent part of the total townscape.

c) Buildings without aesthetic value, which have nothing in common with the general townscape[49].

157
Seventeenth-century patrician houses in the urban sector of the Marais in Paris. The building on the left, which has been neglected for centuries, is now in an advanced state of decay. Extensive conservation work was carried out on the right-hand building five years ago.

158
Façade of a Neo-classical patrician house in Athens after its complete restoration.

49 The mere fact that a particular building does not fit in stylistically with the other buildings in a historic area does not necessarily mean that it should be regarded as 'unaesthetic'. For example, if we find a few *Jugendstil* façades in a medieval urban centre, we should not condemn them out of hand. We have already drawn attention to the unavoidable diversity of structured urban space. None the less, if these individual disparate buildings are to be integrated into the medieval cluster, then their basic urban structure – i.e. the width of their façades, and their overall dimensions – must not be too much at variance with the structure of the medieval centre. Their morphology on the other hand – i.e. their decorative forms, their component materials and the relationship between the wall surfaces and the apertures on their façades – can be completely different and will still not disrupt the townscape.

159
Seventeenth-century houses in Gamla Stan, the old town of Stockholm, which were recently conserved and restored.

4122   The Authenticity of the Form and Architectural Substance of Buildings.
        Patina: Aging with Beauty. The Preservation of Authenticity and Inevitable Interventions
        (Conservation – Restoration – Reconstruction)

Within the general framework of the rehabilitation of historic urban clusters, the protection and conservation of individual buildings – and the spirit in which this work is pursued – plays an extremely important part. Moreover, the restoration and reconstruction of architectural monuments in protected areas and of their accompanying buildings are closely connected with the problems of conservation.

The authentic form and the original substance of buildings are their most important attributes. Unfortunately, these cannot be preserved unchanged for ever and, in point of fact, we find that they are affected by the following factors:

a) Natural aging due to unavoidable wear and tear brought about by variations of temperature, biochemical changes, weathering, air pollution, corrosion (caused by micro-organisms and exposure to the air).

b) Mechanical forces, which disrupt the static equilibrium of a building and may lead to its partial or complete destruction. The principal causes of this kind of damage are earthquakes, inundations, fires and acts of war.

c) Human intervention over the course of the centuries, leading to alterations[50] in the functional arrangement of a building (the organization and use of its internal space) and in its morphological character (the style of the façades and the interior decoration). An extreme form of alteration is the intentional and premature demolition of a building, when this is done simply in order to put the site to a more economic use by erecting a larger building based on modern designs and built by modern methods.

In order to counter these perennial dangers, in order to protect the authentic form and substance of genuine works of architecture and so prolong their lease of life, we have to apply special methods of 'protection' and 'conservation'.

The legal and administrative measures which have been evolved are intended to protect buildings against alteration and destruction as a result of human initiative. But these protective measures, important though they are, do not help with the aesthetic problems posed by interventions. On the contrary, they are concerned simply with the formulation of legal provisions and special building regulations for historic urban centres (see Bibliography 'D').

The natural aging of buildings and the damage done to them by mechanical forces are being countered by contemporary techniques of conservation, restoration and reconstruction. But the deontological problems raised by these techniques – a particular technique and a particular method of intervention has to be decided on in every individual case – are amongst the most difficult and the most crucial in the whole sphere of protection.

We have already given a brief survey of the historical crystallization of the concept of the 'architectural monument' and of the various methods of protection evolved for it over the past two hundred years (see section 21).

Today problems of protection are undoubtedly approached in a strictly scientific manner and the greatest care is taken to ensure that the authentic form and substance of architectural monuments are preserved as far as is humanly possible. By now fortuitous bias has been virtually eliminated and no longer plays any real part in such considerations. And yet our conservationists, architects, archaeologists and art historians are as divided as ever over the way in which essential interventions ought to be carried out.

---

50 The term 'alterations' covers the fortuitous, arbitrary and unaesthetic alterations to which an architectural work is always exposed and not the interesting morphological changes which a building may undergo in the course of the centuries and whose aesthetic appreciation poses one of the most difficult problems within the entire sphere of rehabilitation.

This divergence of opinion is chiefly due to the fact that these specialists tend to lay the main emphasis on one or other of *the two fundamental principles on which the protection of architectural monuments is based. One group insists primarily on the conservation of the authentic form and the original substance of buildings, while the other considers that the real importance of architectural monuments lies in their capacity as urban and historic symbols. This second group is even prepared to accept reconstructions, which are really no more than copies or falsifications of original and authentic structures.*

In the following remarks and observations we shall endeavour to establish – on the basis of concrete examples – suitable criteria for interventions in respect of individual historic monuments and buildings. These will apply to the conservation, restoration and reconstruction of traditional buildings and even their eventual extension in a contemporary style of architecture. We shall also draw attention to the relative nature of the principles involved and to the unavoidable element of subjectivity in this sphere.

We first propose to analyse the question of interventions in respect of individual structures (architectural monuments and their accompanying buildings) and to proceed from there to consider the treatment of groups of buildings in protected areas.

Conservation is the first stage of architectural intervention and may be applied both to living architectural monuments (see 121) and to those which have fallen into disuse.

Conservation work on living buildings, i.e. buildings that are still capable of fulfilling their urban function, includes repair work to make good the worst effects of general wear and tear (painting every five years, replastering and major roof overhaul every ten to twenty years, renewal of floors every ten, twenty or fifty years depending on the type of material). In certain cases, where the structural members of buildings have to be strengthened or repaired, far more basic conservation work will have to be undertaken (replacement of wooden or metal pillars and beams, renovation of vaults cracked or damaged as a result of subsidence, partial renewal of stone foundations which have rotted as a result of physical or chemical processes).

It is widely assumed that this conservation work, which is the first step to be taken in prolonging the life of a building, is not only necessary but generally accepted to be so. But this first intervention also constitutes the first attack on the authentic substance of the building. Consequently, even if the formal authenticity of the building is preserved by careful planning and workmanship – and unfortunately this is not something that can always be guaranteed in advance – this loss of substance is a serious matter.

Over a period of thirty years regular repair work can easily destroy one tenth of the authentic substance of a building, while over a period of one hundred years between ten and twenty per cent of this authentic substance could be lost as a result of structural repairs. (In the case of war damage the percentage may, of course, be very much higher.) In the course of a single century, therefore, the loss of original substance due to unavoidable wear and tear might well amount to twenty-five per cent. And, although the percentage loss fluctuates considerably from building to building, the figures quoted are entirely feasible.

Simple buildings made of costly materials have the best chance of survival. Although the Hephaesteion (Temple of Vulcan), which rises up above the classical agora in Athens, was neglected for fifteen hundred years, the total loss of original substance sustained during that period amounted to no more than twenty-five per cent and was restricted to the timber framing of the roof, the wooden doors, the sheets of marble and the encaustic wall paintings. Buildings such as the patrician houses of Macedonia, on the other hand, have proved far less durable. With their half-timbered walls and their wooden balconies and roofs, these buildings are now in a state of almost total decay due to the lack of systematic renovation.

During the nineteenth century, arbitrary alterations were made to the interior décor of historic buildings in order to bring them more into line with contemporary 'fashion'. This regrettable practice was then extended from the decorative to the structural sphere, where it led to the use of metal constructions as supports for stone vaults and of concrete for the repair of stone pillars (for example, in the Parthenon). Entirely new elements, such as retaining walls and buttresses which radically altered the authentic

160
Nineteenth-century boarding-house in Berlin. The substance of this building is completely decayed.

161
Detail of the outer 'skin' of a Gothic cathedral. Unfortunately, the delicate sculptural decorations on such cathedrals were often counterfeited in the nineteenth century when 'renovation work' frequently involved the production of 'exact replicas'.

architectural form, were also introduced in certain cases, while in the case of the Gothic cathedrals precise but 'dilettante' replicas were made to replace sections of the external décor that had been damaged or destroyed by the effects of weather. During the past few decades, however, far stricter and far more scientific attitudes to conservation have been evolved which, although varying in detail, augur well for the future.

Nowadays, when called upon to renew floors or rafters, or to replace defective stucco and plaster forms, or to renovate paintings and wall decorations, our restorers do their utmost to preserve the spirit of the total composition by studying the architectural plans and any other documents which might throw light on the intentions of the original designer. In the course of their work, they often unearth serious errors perpetrated by their nineteenth-century predecessors.

But, even if the restorers are completely conscientious in their approach, we are still left with the problem of how to avoid introducing new components into a traditional structure, where they must inevitably clash with the older patinated elements and so create a disturbing visual contrast. This aesthetic problem is extraordinarily difficult to solve.

The process of 'artificial aging', in which the new materials used for the renovation of buildings or paintings are specially treated to make them look old, is quite absurd and patently dishonest, for it is diametrically opposed to the whole purpose of renovation, which is to 'renovate' or replace defective parts of an original composition.

However, there is another solution to this problem, which is entirely consistent with the aims of conservation. By using matt colours and unpolished marble for cladding, i.e. by avoiding all bright or shiny surfaces, and by allowing any external materials, such as tiles, to weather naturally before they are used, the restorer can create a relatively homogeneous impression. This is particularly important in respect of roofing tiles, whose form is strongly influenced by their chromatic quality and texture. To patch a roof that has acquired a general patina over a long period of time with brand new tiles produces an optical conflict that is highly disagreeable.

But before continuing with our analysis we must first explain what is meant by the concept of 'patina' and try to define its different aspects.

In the sort of historic building that is a fit subject for conservation, the process of 'natural aging' is accompanied by a parallel process of 'aging with grace', in which a transparent film settles on the external surfaces, thus producing what is known as 'patina'. These two aging processes constantly modify the authentic form and substance of the building, albeit in completely different ways for, while the aesthetic value of the building is decreased by natural aging, it is increased by the patina[51]. (In actual fact, of course, 'natural aging' and 'aging with grace' are so closely interconnected that they really form a single corporate process.)

---

51 Patina increases the aesthetic value of buildings of all ages with the exception of our own and probably those of the future. Contemporary buildings receive no benefit from the effects of patina, but have to be constantly renovated and cleaned. There are two principal points to be noted about this lack of patina in present-day buildings:

a) Contemporary architects are making ever greater use of industrial products such as glass, metal and synthetic materials, which are largely impervious to distortion and have highly polished surfaces with the result that they resist corrosion far better than traditional materials like stone and wood. Moreover, although these industrial products are highly susceptible to the effects of pollution and so dirty very quickly, this in itself is not enough to produce the positive effects of patina.

b) Although contemporary buildings are not 'aging with grace', it must be admitted that there is no real need for them to do so. Both on account of their form and on account of their component materials, these works form such a striking contrast to their natural environment that their introduction into the natural cluster must be regarded more in the light of harmonic confrontation than harmonic integration. Given that this is so, the lack of patina – whose principal purpose is to facilitate the integration of buildings into their natural setting – scarcely matters.

This aging process – especially when it appears on the façades and roofs of buildings – brings out the human element of the urban scene and also helps to integrate the buildings into their natural setting, thus eliminating the visual contrast between man-made structures and the natural environment. But there is also a negative side to this process. An aging building or group of buildings begins by looking desolate, then looks neglected and finally looks like a ruin. Ruins, of course, can be charming. And when we admire the patina of archaeological ruins we do not for one moment think of cleaning their façades or renovating their structure.

The ruins of medieval fortresses are a constant source of fascination, and even historic urban sectors which have reached an advanced stage of decay often give the townscape a picturesque quality.

It would be highly desirable to establish an easily recognizable frontier, a sort of demarcation line between these two types of aging, i.e. 'aging with grace', which is the vehicle of historical memory, and 'natural aging', which is the cause of decay. Unfortunately, it seems scarcely possible. Two disparate trends, both of which appear fully justified in themselves, would have to be reconciled. The first of these, which appeals to the scientific mind, insists on regular conservation in order to preserve as much as possible of the authentic architectural form of historic buildings and centres (a procedure which would, of course, completely eliminate their patina), while the other is more concerned with the preservation of the patina and the 'polished' character of historic architecture and consequently less interested in regular conservation. This impasse may result in absurd solutions such as that of 'artificial aging', which has been described above and which has already been incorporated into various conservation projects. The conservation work undertaken in Paris affords a striking example of the aesthetic problems posed by patina. Because of the great diversity of its townscape, the historic urban centre of Paris, which covered some eight thousand hectares *intra muros* in 1870 and contains architectural monuments dating from the last eight hundred years of French history, is unique in the whole of Europe.

Up to 1963 this historic centre of Paris was covered by a layer of patina which was due to the effects of soot on the worn stones and the crumbling stucco of the façades.

For reasons that were typical of Paris – controlled rents which left the landlords with inadequate means to carry out necessary renovations – this layer of dirt, which had settled on the buildings over the centuries and was several millimetres thick, had never been removed and constituted one of the most characteristic visual elements of the historic townscape of Paris.

But, although this dark veil of patina helped to underline the visual homogeneity of this great metropolis, it destroyed the original chromatic contrast between the yellowish stones of the façades and the dark grey cobbles of the streets and made it difficult to distinguish between different architectural styles.

It was thanks to the initiative of André Malraux, the French Minister for Cultural Affairs, that the façades in the historic centre of Paris were cleaned, for it was he who took the necessary administrative measures. This undertaking is unique in the annals of architectural conservation both on account of its size and on account of the coordination required to carry it through. During the past five years the 'old Parisian skin' has been peeled off to reveal a new city which had never been seen before. The method used – which is known in France as *ravalement* – simply involved washing the façades with powerful jets of water from pressure hoses.

Important complexes such as the Louvre and the symmetrical groups of buildings on the Place de la Concorde can now be seen in all their original glory. The architectural style of different buildings is immediately recognizable and the contrast between different colours and materials is again taking effect. This great project has met with a mixed reception both from the experts and from the public. Some regret the loss of the patina, while others commend this bold intervention which has restored the authentic splendour of an entire city without any loss of architectural substance.

In this particular case we consider that the intervention was a positive act, whose outcome fully justified the sacrifice of the traditional patina. We would not have taken this view, however, if the intervention had not been extended to the whole of the historic nucleus. This is really the crucial point, for if just a few

162
A patrician house in the old Plaka district of Athens. This building, which was erected in 1842, is 'aging with grace'. The patina on its façade has been gathering for the past one hundred and thirty years.

163
The home of the architect Kleanthis in the centre of the Plaka. This house, which was built in 1830, has passed the stage of 'aging with grace' and has now started to decay.

isolated buildings in one sector of the historic nucleus had been cleaned, the contrast between these and the patinated buildings around them would have been both absurd and intolerable. Such aesthetic confusion would have called into question the authenticity of the clean buildings, for it would be difficult to understand why certain buildings from a particular period of the past had been allowed to retain their patina while others from the same period had been completely renovated.

It would also be preferable if the medieval buildings of Paris such as Notre Dame, the Conciergerie and Sainte Chapelle were to be exempted from this general trend towards clean façades. If they were to retain their patina, it would underline their age and establish a desirable contrast between them and the Renaissance, Baroque and nineteenth-century buildings. Unfortunately, it has been decided – after long deliberation as to the feasibility of cleaning the sensitive sculptural detail on these medieval façades – to go ahead with this project.

We have already noted that 'radical conservation' involving structural repairs is now executed in a far less arbitrary manner than it used to be. The materials for such repairs are carefully matched with the original building and, where possible, the freestone for patching walls is obtained from the quarry used by the original architect. *By matching his materials in this way, the contemporary architect engaged in conservation work is able to achieve maximum fidelity for, although his materials are not strictly authentic, their texture is identical with the texture of the original materials.* Having established this essential similarity, however, it is imperative that the architect should distinguish, discreetly but openly, between the new and the original elements of the building in order to preclude all possibility of confusion as to the authenticity of its different components. This can be done in a variety of ways. For example, by shaping his freestone so as to produce 'cutting' edges, the architect can create an effective contrast to the original freestone, whose edges will have been worn away by the effects of the weather.

In the case of timber and metal structures and of stone and tile vaults, the damaged parts are never strengthened by the introduction of structures that are out of keeping with the original architectural design. Rather than do this, contemporary restorers prefer to completely remove and then reconstruct the damaged parts, re-using the authentic material wherever possible. Nowadays it is customary to record the date and nature of such work, listing all parts that have been reconstructed on a small plaque, which is then mounted on a wall in a prominent position.

Reinforced concrete, which is both economical and durable, is also used for conservation work. True, the replacement of a stone or tiled vault by a massive reinforced concrete structure is bound to change the original building. But, on the other hand, the overall monolithic nature of the vault remains essentially the same. Moreover, there is a historical precedent for such structures, namely the *opus incertum* vaults which the Romans cast from solid concrete. This historical precedent, coupled with the undoubted strength of reinforced concrete, must make this material a viable choice for reconstruction work, provided it is used in the right way, i.e. provided the surface of the finished vault remains uncovered. A successful example of a vault reconstructed in reinforced concrete is to be found in the church of St. Julien Le Pauvre on the left bank of the Seine in Paris.

On the other hand, it is quite unpardonable to replace timber beams or framing with reinforced concrete replicas. This lamentable method has often been used in Italy to repair basilicas and other churches whose roofs were destroyed in the two world wars, and was also employed in Greece for the reconstruction of the basilica of St. Demetrius in Salonica, which was gutted by fire in 1917. Since these two materials are entirely different, any attempt to replace timber by concrete must undermine the authenticity of the original substance. It also indicates a complete misconception of the aesthetic and functional values[52] of architectural structures.

---

52 Since concrete and timber have totally different structural characteristics, a framework made of concrete should have a completely different layout from a timber framework. Consequently, any attempt to reproduce a timber framework in concrete must inevitably look out of proportion.

If, for any reason, it is impossible or inadvisable to repair a timber framework with timber so that reinforced concrete has to be used, then the old framework should be completely dismantled and replaced by a new reinforced concrete structure, whose design should conform to the morphological requirements of the new materials. This would then constitute a genuine piece of conservation work, which is preferable to an ungainly and structurally absurd imitation.

The most fragile part of historic buildings is their sculptural décor. This frequently requires renovation, which should be carried out in accordance with one of three tested methods:

a) The replacement of a damaged or missing part by an exact copy executed in a different material. (Example: the plaster copy of the second caryatid on the Erechtheion, which replaced the missing marble original.)

b) The introduction of a new composition executed in the same format and materials and illustrating a similar theme as the missing original but conceived as a contemporary work of art. (Example: The frieze on the restored façade of the Gewandhaus in Braunschweig.)

c) The replacement of a damaged or missing part by a *bas-relief* which is executed in the same material as the original, but which merely indicates its general outline.

In certain cases any one of these methods could be used. Essentially, however, the first – in which the choice of a different material indicates the substitute nature of an otherwise perfect replica – is for free-standing sculptures. By contrast, the other two methods – in which the material is the same but the design is different – should be used for reliefs. The second of these offers the simplest and clearest solution. The first is undoubtedly more creative but is far more difficult to apply. *The one method which should never be used is the replacement of missing sculptures by exact replicas executed in the same material. This solution, which was tried in the nineteenth century (pseudo-gothic forms on cathedral façades) is quite unacceptable since it strives after a false authenticity.*

The intervention techniques outlined above are applied to 'living' architectural monuments in order to protect them and so prolong their life. But there is another form of conservation which, strange as it may seem, is concerned with the protection and preservation of archaeological ruins. How does one preserve a ruin? At first sight the undertaking appears quite senseless, but in point of fact this branch of conservation does exist. Instead of trying to preserve an original functional role (which has long since been lost), the workers in this field merely try to preserve what is left of the original architectural substance and form, which may possess great historical and artistic value.

The most enduring threats to archaeological ruins and monuments have been posed by the weather and by soil subsidence. Of more recent years, these have been augmented by mechanical vibrations (set up by motorized traffic and supersonic aircraft) and corrosion due to air pollution. The conservation techniques employed in this sphere to date have been essentially the same as those used for 'living' architectural monuments. But now entirely new methods are being considered. One of these – the idea of erecting large sheets of synthetic material on light metal scaffolding to form a roof that would protect particularly sensitive archaeological finds against the weather – has already been tried out in Pylos, where the Minoan palace of King Nestor has been protected in this way. But this solution is not an entirely happy one, for it inevitably distorts the architecture and isolates the archaeological site from its natural setting. It has often been suggested that archaeological ruins should be completely covered over by a clinging transparent membrane made of synthetic resin, which would give permanent protection against the effects of weather. But this too would have its disadvantages, for the membrane would inevitably change the original texture of the building materials.

Restoration is the second stage in the preservation of architectural monuments. Unlike conservation, which is concerned with the protection of existing buildings whose authentic form and substance are both largely intact, restoration is devoted to the re-erection of buildings which have been partly or completely destroyed but whose original materials have been saved.

164
Entrance door of a house in the old town of Zürich. The sign above the door records the name of the house (Zum Steg), the year when it was erected (1357) and the year in which it was restored.

165
The Treasury of the Athenians in Delphi. The perfect cut and the texture of the new freestone provides the necessary distinction between the renovated and the authentic parts of the building.

166 and 167
The Roman Temple of Caesar Augustus at Pola which was almost totally destroyed in an air raid during the Second World War. When it was restored after the war, the original materials were re-used.

The legitimacy and desirability of conservation are axiomatic. But this is not the case with restoration. Indeed, the aesthetic quality of much of the restoration work carried out during the past two hundred years is extremely doubtful.

It is all the more necessary therefore that we should establish precise, objective criteria for interventions of this kind.

The first attempts at restoration were prompted by archaeological monuments. In the course of excavations, important and authentic materials were discovered near architectural ruins, while individual parts of the original buildings were found scattered about in the immediate vicinity. The idea was conceived of reconstructing these parts of buildings, using authentic materials so far as these were available. This idea has remained the basis of all restoration projects. It follows, therefore, that restoration should be restricted to re-erection on their original site of authentic parts of historic buildings. New building materials may be introduced if absolutely essential, in other words if they are needed to ensure structural stability. Where they are introduced, however, techniques similar to those used in conservation work must be adopted in order to distinguish between the old and new sections of the structure.

Because of their great age archaeological monuments appeal strongly to the imagination, even when they are in ruins. Consequently, it is not necessary to reconstitute a whole building in order to bring out its symbolic and historical significance. The theory that the symbolic power and psychological appeal of archaeological ruins would be enhanced by comprehensive restoration is completely false and merely testifies to a dearth of historical imagination. This kind of approach inevitably leads to the reconstruction of pseudo-antiquities, a practice which currently goes under the name of the 'creative care of monuments' and which is totally unacceptable since it constitutes a conscious falsification of archaeological monuments. It could, of course, be countered that there is no such thing as complete authenticity in respect of the material composition of monuments. But this argument is essentially specious for, although it is unavoidable that some part of the authentic substance of a building should be lost as a result of the conservation and renovation of 'living' architectural monuments, this loss is fully justified by the obvious need to prolong the life of such monuments. This does not apply, however, to the large scale reconstruction of archaeological ruins, which is based on a purely arbitrary decision.

But, quite apart from archaeological monuments, we also have to consider the question of the restoration of 'living' buildings, which formed an integral part of the urban image before they were destroyed (by acts of God or acts of war). In this sphere there are further cogent reasons why restoration should be undertaken. These are:

a) The fact that these buildings were inhabited and fulfilled specific functions prior to their destruction.

b) The importance of such buildings as part of the urban cluster and their contribution to the homogeneity of the townscape.

c) The symbolic function of such buildings in urban space. (This applies primarily to churches, architectural monuments etc.)

d) The sentimental attachment of the population to such buildings (which had always formed part of the urban scene).

These were the principal reasons underlying the restoration of so many buildings of aesthetic and historical value, destroyed in the two World Wars.

In these projects, the authentic material found near the damaged buildings was mostly rejected because it was found to be no longer suitable. (Many of the stones and bricks had been weakened by the effect of the heat and a great deal of the wood had been charred.) Instead, new materials were used and integrated into the buildings in the approved manner (as described above in the section on conservation). Moreover, immense care was taken to ensure that the design of these reconstructed buildings matched that of the originals.

Although it is extremely regrettable that so much of the authentic substance of these buildings should have been lost in the course of restoration, this was perhaps acceptable in so far as they were living buildings and not archaeological monuments. None the less, it would be a serious error to assume that these methods of restoration may be automatically applied in each and every case. There is considerable doubt as to their legitimacy, especially where they involve a loss of over fifty per cent of the authentic substance.

In certain instances, of course, the symbolic importance of a monument may be so great that when it is destroyed, it is immediately restored, even though only a small percentage of the original materials have survived. A striking example of such restoration or, to be more precise, reconstruction is afforded by the Campanile of St. Mark's in Venice which was totally destroyed in the early years of the century and immediately rebuilt in its present form. But even in such special cases, it would be preferable to find other more genuine and more consistent ways of preserving the historical and symbolic value of a destroyed building. Complete reconstruction is, after all, a falsification. One brilliant solution to this problem was found by the German architect, Egon Eiermann, in his conversion of the Kaiser-Wilhelm-Gedächtniskirche in Berlin. In this project, Eiermann demonstrated beyond all doubt the extraordinary contribution which a contemporary ruin can make to the townscape and the psychological impact it can have on the popution. His Gedächtniskirche is a salutary reminder that it is often better not to try to reconstruct completely or badly damaged historic buildings.

Where buildings have been only partially destroyed, another method of restoration is possible. This consists of adding new wings or storeys designed in a contemporary style to the surviving section of the original structure. This method does, of course, involve a tremendous stylistic contrast. But there are many precedents in the history of architecture, including the celebrated Doge's Palace in Venice, whose façade is a perfect example of the harmonic integration of three different styles, namely Romanesque, Gothic and Renaissance. Morphological diversity in a building can be both interesting and attractive (see 4125).

But if this method of restoration is to be successful, there must be an organic link between the old and new sections of the building. The best way of achieving this is by designing the extension in the form of a wing. This gives a horizontal layout which, if successful, will create an impression of harmonic contrast between the different sections of the building. (See 4123 for a description of 'harmonic contrast' between the different buildings in a historic centre.) The organic link between the old and the new structures is not likely to be achieved, however, either by vertical extension (adding further storeys executed in a completely different style to the surviving lower storeys of an existing building) or by incorporating the surviving sections of a damaged building into an entirely new structure. All that this is likely to produce is dissonance and morphological confusion.

If an architectural monument is completely, or almost completely, destroyed it cannot be restored. What often happens in such cases – provided the foundations and main structural members still exist and the original architectural plans have been preserved – is that the authentic building is replaced by an exact copy, a counterfeit. Such reconstruction projects serve no positive purpose. The reproductive processes employed in other spheres of the fine arts (namely painting and sculpture) may appear dubious when considered in ethical terms, but they do at least serve a positive purpose by providing wider access to works of art and thus educating the public in this particular sphere. *This is not the case with architectural reproductions. Consequently there is no justification for a counterfeit building, which can only possibly appeal to naive and romantic people, who have no real understanding of art and who apparently prefer to acquire their knowledge of architectural history from these life-size models.*

168 and 169
For the restoration of the principal façade of the Braunschweig Gewandhaus, entirely new sculptures were commissionde. These were made from the same material and were of the same size as the originals.

170 and 171
Nave of the Basilica of San Lorenzo fuori le Mura after its destruction in the Second World War (170) and during restoration (171). The timber frame of the roof was reproduced to the same scale and in the same material.

The Stoa of Attalus on the eastern side of the classical agora in Athens, which was reconstructed between 1956 and 1960. We are indebted for this admirable piece of reconstruction work to John Travlos, the Greek architect and student of Athenian history. But, remarkable though his achievement was, this project was not really justified. Travlos did not restore an archaeological monument; he took a reconstructed model and reproduced an architectural model on a natural scale.

173

An aerial view of the sanctuary at Delphi. In this sanctuary only the Treasury of the Athenians, the Stoa of the Athenians and the pillars of the Temple of Apollo have been restored. But, although limited, the restoration work done in Delphi has been carried out with great care and only authentic materials from the original structures have been used. This kind of partial restoration is highly effective. It provides the observer with certain basic points of orientation, and then leaves it to him to complete the townscape by the application of his historical knowledge and imagination.

174 and 175
A striking example of the liberties taken by nineteenth-century restorers. The almost ruined castle of Pierrefonds in France before (174) and after (175) its 'restoration'. This project was undertaken by the celebrated French architect and restorer, Viollet-le-Duc, at the command of Napoleon III.

We must now turn from interventions in respect of individual buildings in protected areas, to interventions in whole groups of buildings. Because so many of these groups have been so badly damaged in our time, due both to the devastation of war and to urban structural changes, they pose a very special problem. Consequently, it must be said at the outset that although reconstruction has been rejected as an acceptable procedure in respect of individual buildings, it cannot be dismissed out of hand where groups of buildings are concerned. Because of its special importance, this question will be dealt with in detail on a later page (see 4123).

As far as conservation and restoration is concerned, groups of buildings are treated in much the same way as individual buildings. This procedure has, of course, already been described in full. But there is one point which deserves special consideration, namely the effect which the 'patina' of groups of buildings can exert on the townscape.

There is no set timetable for the conservation and restoration work carried out on individual buildings in urban areas. The local authorities or house-owners have this work done to suit themselves. As a result, we often find a marked contrast in the appearance of different buildings within the same group. Now, although the plurality of the townscape is entirely acceptable when it is caused by the juxtaposition of buildings dating from different periods and designed in fundamentally different styles, it is unacceptable when due to a lack of coordination in the renovation of buildings from the same period. The juxtaposition of a group of patinated buildings with a group of newly renovated buildings from the same period reduces the aesthetic appeal of a street or district and should be avoided at all costs. The buildings which are covered with patina then look neglected and create a dreary impression when they are in fact 'aging with grace'. The newly renovated buildings, on the other hand, look 'more than authentic', as if they were perfect replicas rather than genuine historic buildings.

It is highly desirable, therefore, that the conservation and restoration work on buildings of the same group which date from the same period should be undertaken at the same time. The coordination of such work is naturally a matter for the local and national authorities.

176 to 181
Living architectural monuments in Italy which were bombed in the Second World War and subsequently restored.

176   The Palazzo della Mercanzia in Bologna (built 1384) after an air raid in 1943.

177 and 178   The Piazza San Carlo in Turin after the raids (177) and after its restoration (178).

179, 180 and 181   The Basilica of San Lorenzo fuori le mura in Rome with its principal façade and portico completely destroyed (179), during the restoration work (180) and before the war (181).

The Breitscheidplatz in West Berlin with the Kaiser-Wilhelm-Gedächtniskirche in the centre. This church, which was almost completely destroyed in the Second World War, could have been effectively restored, since most of the original building materials were still to hand. It is fortunate that this course was not taken, for the Gedächtniskirche is a pseudo-Romanesque construction erected in the closing years of the nineteenthcentury-which means that we would have been confronted with a counterfeit of a counterfeit. Instead Egon Eiermann's scheme was accepted, and the gutted and battle-scarred tower of the original building, after being adequately strengthened, became the central feature of a new structure, in which it is flanked by the six-sided nave and the belfry of a contemporary church. By rejecting restoration and integrating a 'ruin of our own times' into a completely new building, Eiermann created a powerful urban symbol which testifies to both the suffering and the rebirth of the city.

183

View by night of the monumental new Gedächtniskirche. Photographed from the east.

Reconstruction within the Sphere of Historic Centres. Scenographic Solutions.
Harmonic Integration and Contrast

Following the vast devastation wrought by the Second World War in so many historic towns (Warsaw, Hanover, Berlin, Munich, Dresden, Coventry, Florence), the architects and specialists in conservation, the art historians and the members of the general public were obliged to come to terms with the problems of reconstruction. The aesthetic demands made on this post-war generation were very considerable, for they found themselves faced with the task of re-erecting whole groups of buildings and, in some cases, whole urban areas within their historic centres.

There were two principal reasons why this reconstruction work had to be undertaken. In the first place the commercial, cultural and administrative centres of these major cities had to be revived, and in the second place the symbolic significance of their historic centres had to be preserved, especially in view of the strong attachment shown for them by the local inhabitants.

After the Second World War numerous individual buildings were demolished, both by private builders and by the local authorities in the course of reconstruction programmes and new sanitation projects. (Where these buildings were situated in protected areas, the greatest possible care was taken over the demolition work.) In all cases the object of the exercise was to create entirely new structures which would make fuller use of the original sites. In this connection, it should be pointed out that the reconstruction of groups of buildings in protected areas is usually unavoidable since the legal and administrative protection of historic settlements is not designed simply to keep the status quo. On the contrary, it provides for the renewal of buildings in these settlements, which is often highly desirable.

In the course of this reconstruction work, a number of different trends have come to light.

The great mass of the population and many intellectuals who are entirely conservative in their approach[53] and who, for purely sentimental reasons, consider that only the symbolic and historical role of buildings is at all important, favour the 'restoration' of their historic centres (although in most cases during the post-war period this has been tantamount to 'reconstruction' since nearly all of the original substance was missing). National and regional authorities and study groups responsible for the conservation work carried out on historic buildings have often tried to implement projects of this kind and their work has led to the development of two more or less conservative trends. These are:

a) 'Historicizing' reconstruction, in which perfect but counterfeit replicas of historic buildings are produced.

b) 'Harmonic integration', in which the demolished buildings are replaced by buildings designed as contemporary structures but in the same spirit as the traditional buildings.

Many town-planners and younger architects, on the other hand, are in favour of introducing contemporary buildings in the proper sense into protected areas. They are firmly convinced that the fortuitous destruction of large areas of historic settlements offers a unique opportunity for completely restructuring these urban centres and renewing their buildings. This attitude has led to the development of a third trend, namely:

c) 'Harmonic contrast', in which reconstruction is based on contemporary principles of town-planning and contemporary architectural morphology.

184
The Leinenschloss in Hanover after its conversion into the Lower Saxony Parliament building. Note the contemporary wing used to replace part of the historic building destroyed as a result of air attacks in 1943. (Conversion carried out between 1957 and 1967. Architect: Oesterlen.)

185
The Pellerhaus in Nürnberg following its postwar restoration and partial reconstruction. Here new storeys executed in a contemporary style have been added on to the base of a patrician house of the Baroque period. (Architects: F. and W. Mayer.)

53 The conservative attitude of the great mass of the population *vis-à-vis* new architectural forms poses a perennial problem for the town-planners. Its explanation is simple: the houses and the towns in which people grow up constitute the most intimate sphere of their lives. Consequently powerful psychological resistances have to be removed before they can grow accustomed to radical change or renewal in this sphere and learn to accept it.

186 to 190

The small medieval fortress of Godesberg in Bad Godesberg on the Rhine after its restoration and conversion into a hotel. Here too new wings designed in a contemporary style have been added to the surviving and restored parts of a historic building. In this particular case the architect, Gottfried Boehm, has been extremely successful. He has followed the general design of the original fortress, enhanced the value of its restored components by giving them new functions, and added contemporary sections, whose formal severity and surface texture enable them to blend perfectly with the historic remains.

186, 187 and 188   Details photographed from the inner courtyard.

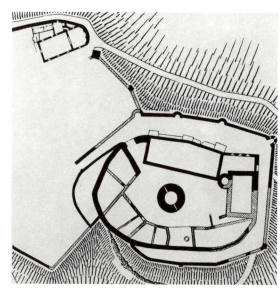

189   The fortress with the new wing which contains the restaurant.

190   General plan of Godesberg.

191 and 192

The Garküchenplatz near the cathedral in Frankfurt before the Second World War (191) and after its postwar reconstruction (192). This is an example of bad reconstruction. The architect has completely failed in his attempt to achieve the harmonic integration of new buildings into a historic setting.

Although the first of the two conservative trends, i.e. 'historicizing reconstruction', provides the broad mass of the population with the comforting impression that the aesthetic climate and the traditional framework of their trusted townscape have been re-established, it actually constitutes a lamentable falsification[54] which any aesthetically-minded person is bound to reject. The arguments against the historicizing reconstruction of individual buildings (see 4122) are equally applicable here.

But in the case of the second conservative trend, which advocates the demolition of the damaged buildings and their replacement by entirely new structures conceived in the morphological 'spirit' of the historic centre, the charge of 'structural falsification' cannot be upheld.

---

54 In the case of historic buildings, whose stone walls survived destruction but whose ceiling and floor timbers were burnt out, historicizing restoration is not a falsification. Moreover, a project of this kind would give an opportunity of introducing up-to-date sanitation, which means that it should really come under the heading of necessary restoration and not reconstruction.

193
Three-storeyed apartment house built in 1960 and flanked by nineteenth-century buildings in Mannheim. This attempt to achieve harmonic contrast is entirely successful. (Architect: C. Mutschler.)

194
Principal façades of new buildings near the Ponte Vecchio in the centre of Florence. Despite the immense difficulties posed by the site, this attempt at harmonic integration is undoubtedly successful.

195
Pseudo-Hanseatic burghers' houses built in Danzig before the Second World War. This is a typical example of scenographic reconstruction, which is invariably prompted by a misguided desire to re-establish the symbolic importance of an architectural monument.

196 and 197

Two views of groups of reconstructed buildings near the Ponte Vecchio, whose principal façades overlook the Arno. The original buildings, which were erected during the Tuscan Renaissance, were blown up during the German retreat in 1944. In our view this large scale reconstruction project is the most successful example of harmonic integration in the whole of Europe.

198

Harmonic integration in the historic urban centre of Hanover.

The use of up-to-date building materials and the functional organization of the rooms testify to the contemporary character of these new buildings. As for the architectural detail and general form of their façades (width of the buildings, shape of the roofs, projecting cornices and balconies), here a compromise solution has been reached. Wherever possible the old traditional forms have been preserved with the result that, by and large, we find that the width of the façades is the same, that the original ratios between the wall areas and the door and window apertures have been preserved, and that the arrangement and size of the projecting structures and the choice of colours have remained unchanged. This fidelity *vis-à-vis* traditional values has had a fortunate outcome, for the buildings conceived in the morphological spirit of historic centres fit in with their surroundings extremely well. None the less, no attempt has been made to create perfect replicas. New building materials are invariably used (unclad concrete and cement rendering instead of the traditional stone blocks) and the moulded or painted decorations, which are a feature of the original buildings, are never reproduced, although in some cases they are replaced by new decorative compositions executed in a contemporary style. This second conservative trend constitutes a genuine attempt to achieve 'harmonic integration' without introducing falsifications into the historic settlement.

The third trend that has been evolved for the reconstruction of buildings and groups of buildings in historic centres is aesthetically the most rewarding. Its object – one that has often been pursued in the historical development of architecture – is 'harmonic contrast'. Basically, this consists of replacing derelict buildings in protected areas by completely new structures conceived in contemporary styles, which will then set up an extreme but 'harmonic' contrast with the existing historic buildings. *The layman is both shocked and frightened by this idea because he is unable to accept that historic building complexes should be replaced by 'modern' buildings.* But we need only consider the architectural changes wrought by the creative spirit of the Renaissance in the historic centres of the Middle Ages to realize that, *far from being offensive or outrageous, the bold juxtaposition of different architectural styles is based on extremely reputable precedents in the historical development of architecture and town-planning.*

But what is 'harmonic contrast'? In the first place, this contrast is produced by the juxtaposition of highly divergent architectural styles. For example, an extreme contrast is produced if a building with a glass front or an unclad reinforced concrete framework is erected in the immediate vicinity of medieval buildings constructed from stone blocks, which reveal highly finished terraced façades with narrow doors and windows and rich ornamentation. However, this contrast is not visually offensive. On the contrary, it has a very positive psychological effect because it enables the viewer to compare the artistic achievements of different epochs. *But this contrast will only be harmonious if the scale of the new buildings fits in with the scale of the buildings in the historic area.* The materials used for these new buildings, their disposition and function should reveal no trace of imitation. On the other hand, their designers should respect the scale of the urban setting and the height of the buildings, their proportions and layout ought to conform to the general scale of the historic centre, thus ensuring the harmonic integration of these new structures into the traditional urban area.

There are, of course, considerable differences of scale between the various kinds of historic settlements. In the medieval historic centres, for example, which are characterized by the free alignment and great diversity of their buildings and by their dense networks of narrow winding roads, everything is on an intimate scale. If this scale is to be reproduced in the reconstruction projects, then – even if the principle of harmonic contrast is adopted – the new buildings will have to be approximately the same size as the old ones and will have to be sited on the traditional road network.

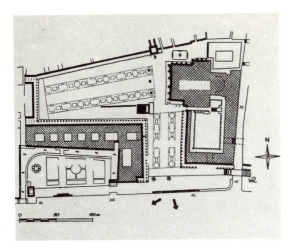

199 and 200
St. Mark's Square in Venice: the most important historic example of the harmonic confrontation of buildings from different periods. Despite the wide variations of style in these closely packed buildings, the square creates a homogeneous impression. Scale of platz 1:5,000.

201 and 202
The principle of harmonic contrast also determined the composition of the main façade of the Haus der Bürgerschaft (Parliament building) on the market place in Bremen. By breaking this façade down into eight equal sections and by paying careful attention to its scale, the architect, Wassili Luckhardt, was able to integrate his contemporary building into this historic square.

201 View from the market place: on the left the Town Hall, in the centre the Cathedral, and on the right the New Parliament building.

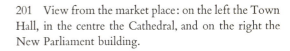
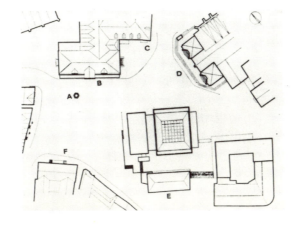

202 Plan of the market place in Bremen: (a) Roland-säule [Roland column]; (b) and (c) Town Hall; (d) Cathedral; (e) The new Parliament building. Approximate scale 1:3,000.

203 to 209

New History Museum, Am hohen Ufer, Hanover (built between 1963 and 1967). Although this new museum is an important building which occupies a complete block, its architect, Oesterlen, has none the less succeeded in integrating it into the old town of Hanover. He achieved this partly by grouping his buildings around a central courtyard – a particularly happy solution – and partly by the extremely subtle partitioning of their façades. Over and above this, Oesterlen was also able to incorporate the remains of the medieval city wall and the Beginnentuim into his museum without destroying their authenticity.

203   The historic remains before the museum was built.

204   The western façade of the museum. This façade has been integrated with the historic remains. The massive projecting roof of the new building, which protects the medieval wall without touching it, is a very ingenious solution.

205   The southern façade of the museum.

206   Plan of the Museum am Leineufer and a view of the half-timbered medieval houses facing it.

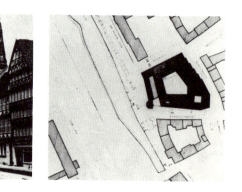

207   The eastern façade of the museum. The dynamic structures created by the partitioning of the façade establish a visual bond with the projecting façades of the half-timbered medieval houses on the opposite side of the road.

208   Another view of the southern façade.

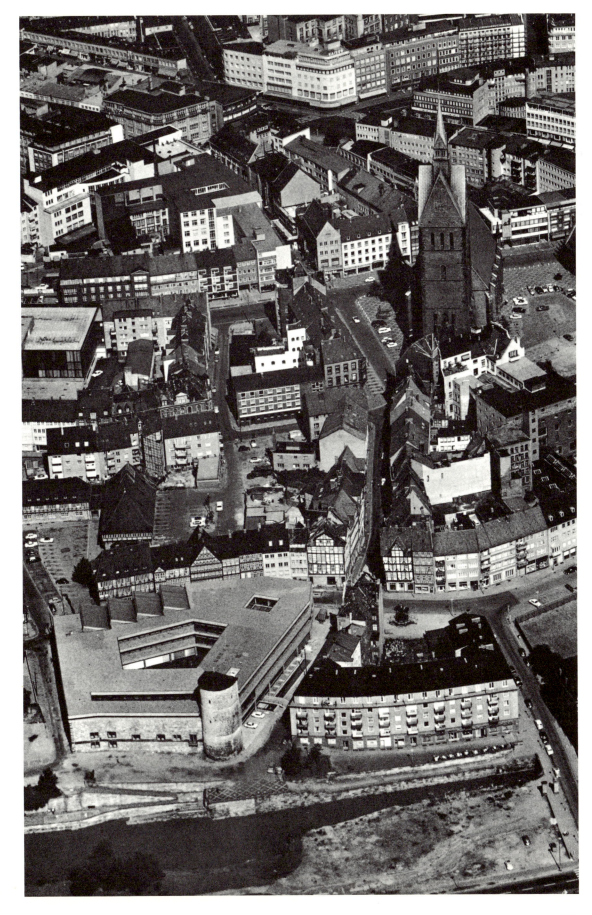

209 Aerial photograph of the historic urban centre of Hanover. The new museum is in the central and bottom sections of the picture.

210 and 211
New house under construction (1968) in the old town of Zürich on the banks of the Limmat. It is quite obvious (fig. 211) that great care has been taken to keep to the architectural scale of the district and to maintain the traditional ratios between wall areas and apertures.

212
Similar examples of harmonic integration in the historic urban centre of Hanover.

213 to 216
The thirteenth-century Salomon Tower in Visegard, Hungary. This is an extremely interesting example of a restoration project based on the principle of harmonic contrast. Note the difference between the old and the new materials.

213 View of the restored tower.

214 View of the ceiling. The original groin vaults have been replaced by flat concrete ceilings. The light metal gratings fixed beneath them indicate the original form of the vaults.

215 The missing part of the tower was reconstructed in unclad concrete, thus creating the necessary contrast with the adjacent original brickwork.

216 Although the new spiral staircase is made of reinforced concrete, which is a contemporary material, its construction is identical to that of the medieval original.

217 and 218

The most impressive Romanesque ruin in Germany is to be found in Bad Hersfeld. Frei Otto had long advocated that the nave of this ruined church should be covered with an awning (made of synthetic materials and supported by cables attached to a steel mast) and in 1968 his proposal was finally put into effect. While any other type of covering would have undermined the authenticity of the ruins, this light and revolutionary structure – which recalls the protective awnings of the Roman theatres and amphitheatres – does not, since it avoids direct contact with them.

219 to 221

Scheme put forward by the German architect Neumann for a students' hostel in the old town of Regensburg in Southern Germany. Neumann proposed to convert the old buildings in the heart of the historic centre (marked with bold outlines on the plans) and integrate them with new buildings with highly irregular forms and of different heights, thus enabling these traditional structures to fulfil new functions.

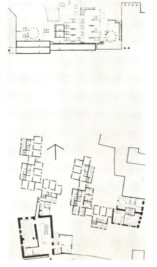

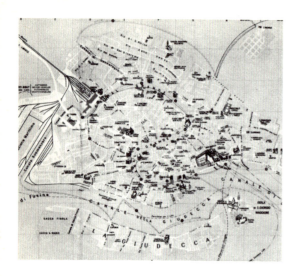

222
Plan of Venice.

223
View of St. Mark's Square and the Doge's Palace from the sea.

224
Model of Le Corbusier's last project (1964): a hospital with 1200 beds planned for Venice. It is noticeable that, in this work, Le Corbusier dispensed with the massive vertical components whose violently sculptured forms were the dominant feature of his late style, and created a basically horizontal design in which he put the traditional forms of Venetian urban architecture – i.e. the piazza (a small square providing a variety of perspective views) and the 'stilt foundation' – to new use. This design is an extremely successful example of harmonic contrast. It is also an object lesson in the integration of a new building into the inner nucleus of a historic area.

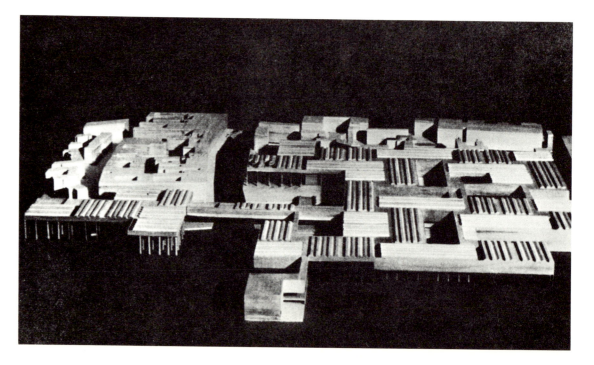

In the historic settlements dating from the eighteenth and nineteenth centuries, on the other hand, the road networks are far more regular and the residential buildings occupy far larger sites. Consequently, where whole groups of buildings have to be reconstructed, it is possible to introduce entirely new principles based on present-day town-planning schemes. In such cases extreme contrasts can be achieved not only by the contemporary styling of the individual reconstructed buildings, but also by their realignment within urban space. Instead of the house façades being set out in rows along either side of the road as in the traditional scheme, they can be planned as free-standing buildings. But here, too, harmony is assured provided the scale of the new buildings is related to the scale of the traditional urban centre.

In certain isolated cases, where new buildings have to be erected on important squares or intersections, it would be preferable if they were set out in exactly the same way as the traditional buildings which they are replacing, not in order to conform to the dictates of 'local colour' by repeating the line of the old façades, but in order to make absolutely sure that in these special areas the scale of the traditional composition is not jeopardized in any way. A very successful example of this kind of project is provided by the UNESCO building on the crescent-shaped Place Fontenoy on the axis of the Champs de Mars in Paris.

The first of the three trends that we have mentioned, namely 'historicizing reconstruction', appears completely misguided. But the other two – 'harmonic integration' and 'harmonic contrast' – are both aesthetically and psychologically acceptable since they are not intended to produce a false impression of age. The choice as to which of these two methods should be adopted depends in the final analysis on the individual architect concerned, although public opinion is also a significant factor in many cases.

225 and 226
Two further examples of clearly defined harmonic contrast.

225 The eleventh-century Byzantine church of Kapyikaréa in Athens, projected on the plain façade of a neighbouring contemporary building.

226 The old central building of the nineteenth-century Technical University in Berlin, projected on the new twelve-storeyed wing that was erected immediately behind it in 1967.

A ridiculous example of translocation: the façade of a burgher's house executed in the highly ornamental style of Hanseatic Rococo was removed from its original site and used to provide a scenographic façade for the new savings bank erected on the market place in Bremen in 1957.

227 The Rococo façade on its original site on the banks of the Weser.

228 The same façade after its removal and re-erection on the market place, two kilometres away from its original site.

4124   The Originality of the Site. Translocations

Although authenticity of form and substance are the most significant attributes of an architectural monument, the authenticity of its site is also an important factor.

Of recent years our conservation, restoration and reconstruction projects have been complemented by 'translocations', which involve the removal and re-erection on a different site of whole façades and, in some cases, of whole buildings. Many of the structures moved in this way were taken from one site within a protected area to another, while others were brought into protected areas from outside.

The principal reason why such operations were mounted was to 'fill up' gaps which had been created in the historic centres, using completely 'authentic' materials. Thus, instead of reconstructing devastated buildings in one of the three ways outlined in the preceding section, authentic buildings dating from the same period and executed in the same style were used as replacements.

Translocations are also undertaken for other reasons. For example, when valuable groups of historic buildings impede an urban development project which is necessary for the rehabilitation of a historic centre (the building of access roads, the installation of parking lots, the creation of 'perspective views' of the focal points of interest of the urban composition etc.), recourse may be had to translocation in a naive attempt to facilitate the intervention while at the same time 'saving' the threatened buildings. But such motivations cannot really justify translocation or atone for the harm done to the townscape in this way. For this technique does violence to one of the basic attributes of an architectural work, the authenticity of its position within the urban cluster. *This authenticity is not just a vague concept but a concrete reality, for when a building is arbitrarily transferred to another site, it loses its whole raison d'être, which is indissolubly linked with its functional and aesthetic relationship to its original urban environment.* The psychological effect which a historic building exerts on those who see it daily also depends to a very large extent on its site, for such buildings are not meant to be mobile. On the contrary, their static nature is fully accepted by one and all. Consequently, all translocations must inevitably give the observer a great psychological shock. The aesthetic value of a building is automatically undermined if its authentic site within the urban cluster – which its architect had intended to be permanent – is not respected.

There is a close parallel to the translocation of buildings, which effectively illustrates the problematical nature of such undertakings. Every human being has a completely individual face, which – apart from aging – remains permanently the same throughout the whole of his life. On the other hand he is able to move about freely in space. Consequently, although it is perfectly normal for a human being to change his location, any wilful alteration to his physiognomy (plastic surgery undertaken for aesthetic purposes) is completely abnormal and constitutes an affront to the authenticity of its original form. This is doubly true of historic buildings which are characterized both by an authentic and original form and by an authentic and unalterable site. Consequently, if either of these characteristic attributes is interfered with, the aesthetic value of the building concerned is automatically reduced; and since alterations to the authentic form of a building are unfortunately inevitable (due to natural aging), it is all the more desirable that its original site should be preserved. The translocation of façades or, for that matter, of whole buildings serves no useful purpose and should be discontinued.

Quite apart from the harm done by the actual translocation of a building, much of its original substance and structure is lost as a result of such operations. Unless a building is designed as a mobile construction, it simply is not possible to demolish it and then re-erect it on a new site without destroying important parts of the original substance (mortar, rendering, stucco, tiles, murals, ceilings, etc.).

But the loss of the authentic structure is even more serious. In order to demolish, transport and re-erect a building, present-day techniques are employed which are completely at variance with the original constructional method.

Moreover, the influence exerted on the population by a translocated building will be minimal as compared with the influence which is wielded from its original site. Even if a building has been damaged, it will create a far greater impression as a ruin on its authentic site than as a reconstructed building on a different site. When a historic building suddenly appears in a new setting, it looks out of place. There is something strange and inexplicable about it.

It is quite evident, therefore, that where architecture is still static (where it is still based on craftsmanship and not on industrialized processes), and where the townscape is still traditional (where it has not yet become flexible and mobile), the translocation of buildings is a completely arbitrary and absurd scenographic solution, which is aesthetically unacceptable and psychologically disturbing.

4125   The Formal Diversity of Individual Buildings. The Accumulation of Architectural Substances

We have already considered the problem posed by the juxtaposition within the urban cluster of various groups of buildings dating from different epochs (see 33). But morphological diversity is not only a characteristic of the townscape. It is also encountered in individual buildings.

Over the centuries the façades of numerous buildings have been repeatedly modified, with the result that they now bear witness to many different styles. The addition of new wings or storeys, the renewal of painted or sculptural decorations and extensions or alterations to courtyards and gardens have inevitably altered the original architectural form of such buildings.

This in turn has led to the evolution of two equally justified but diametrically opposed trends which make themselves felt in the sphere of conservation and sanitation (see 423). One of these trends is prompted by the understandable desire to preserve as far as possible the external signs of the historical development undergone by traditional buildings, which means that mutually alien features are often retained, thus ensuring the maintenance of buildings which lack morphological homogeneity. The second trend is based on the idea that historic buildings, to which alterations and additions have been made in the course of time, should be restored so as to reflect as far as possible the intentions of their original architect.

But although these two trends could scarcely be more disparate – one attaches importance to the historical development, the other to the original condition of traditional buildings – it is not really possible to subscribe unconditionally to either one of them. The truth of the matter is that each is applicable in particular circumstances. The second technique, which involves 'stripping' the buildings, should be used when the later additions are inferior to the original core or, alternatively, when these additions, although aesthetically and historically valuable in themselves, disrupt the unity and purity of the authentic composition to an excessive degree.

229   View of one of the side façades of the new savings bank, which forms a grotesque contrast to the counterfeit *mise-en-scène* of the principal façade.

230
View of the Acropolis in Athens based on a lithograph by Ferdinand Stadtemann. The Frankenturm [Franconian tower], which was demolished in 1875 at Schliemann's expense, is still plainly visible. Clearly, the removal of this medieval building restored the classical composition to its original condition. But it also destroyed the morphological diversity of the site, which had been the outcome of a natural and centuries-old development.

231
Formal diversity in one and the same building. An investigation was carried out in Stockholm in order to identify the different architectural layers of this old burgher's house. In diagrams A and B the black areas represent medieval components, the grey areas seventeenth-century components; diagram C shows three phases in the development of a gable (1300, 1625 and 1650); diagram D shows two phases in the interior layout of the first storey of a burgher's house.

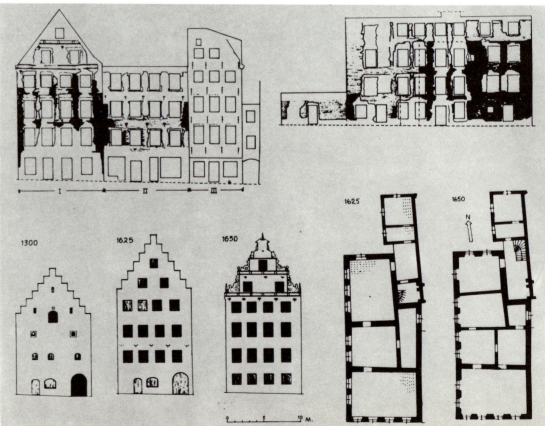

If, on the other hand, a building has been created by a gradual process of evolution and achieves its final form only at the end of decades and, in some cases, centuries of work, then its morphological diversity is an inherent condition and not the result of arbitrary alterations. An example of this kind of evolutionary process is furnished by the Romanesque cathedrals of Western Europe, many of which took three or four centuries to build and consequently contain Gothic as well as Romanesque features (pointed arches or Gothic groin vaults and flying buttresses as developments of the tunnel vault and the plain and massive Romanesque pillars).

In such circumstances it would be inconceivable to suggest that a building should be 'stripped', since this would be tantamount to destroying part of the building itself. Consequently only the first of the two techniques, that of conservation, should be applied here. This would ensure that all the different elements in the building received due and equal attention, even if they were morphologically diametrically opposed to one another. But this kind of project is seldom simple. Although, as in the case of the Romanesque cathedrals, the accumulation of architectural forms and substance might be due to the introduction of new elements which remained relatively independent of the older structures, in the majority of cases morphological plurality results from the superimposition of new elements on one and the same façade. When this happens, success depends entirely on the experience and sensitivity of the architect in charge of the project, for he will have to pick out what he considers to be the most valuable of the successive phases in the development of the building and emphasize this at the expense of the others. In making his choice, he will be guided, not by the value of the original architectural form, but by the aesthetic and historical appeal of the various evolutionary phases through which the building has passed and by their affinity with the general townscape.

The problems posed by the formal diversity of historic buildings and the accumulation of architectural forms and substance are considerable, and the decisions taken in this sphere are highly susceptible to criticism since they are largely based on subjective responses. For this reason also, it would be useless to try to establish rules or criteria for determining which evolutionary phase is to be stressed and which phases are to be eliminated.

Today there is a growing tendency to favour the 'stripping' of historic buildings and although the success of such projects depends on the 'subjective sensitivity' of a single expert, this technique has often been applied successfully and has led to the revival of many valuable buildings.

232 and 233
Seventeenth-century patrician house in the urban sector of the Marais in Paris, which originally belonged to the Count of Sully. In the early years of the twentieth century, this building was defaced by an incongruous addition and by unaesthetic commercial signs (232). Today, after its total rehabilitation, it is once again a gracious and highly attractive example of French Baroque architecture (233).

421   *The Incorporation of Present and Future Life into Preserved Areas.*
        The Living Conditions and Needs of the Population and their Satisfaction

If the aesthetic problems discussed in the preceding sections of this book (see 411 and 412) are dealt with at the right time and in the right way, the protection of the traditional townscape of historic centres will be more or less guaranteed. But, although they will certainly ensure the physical survival of urban compositions, these measures cannot possibly preserve the social life of our historic centres. This calls for more far-reaching measures.

The principal task here is to preserve the uninterrupted flow of 'urban activity' in our historic centres and make it possible for social life to continue (see 121). For if these 'urban events' were to decline or die out entirely, then such settlements would soon be transformed into deserted ruins. The most they could hope for then would be to be rediscovered at some future date as archaeological sites[55].

In our efforts to preserve an active social life in our historic settlements, we are confronted with a strange situation. *We find ourselves faced with an 'object' of historical origin which 'contains' human activities and which, although protected on account of its aesthetic value, is not confined in a museum. On the contrary, because of its architectural nature, it still occupies its original place and largely maintains its original function. But, although it is certainly to be hoped that this 'vehicle of human activities' will continue to exercise this function in the urban cluster of the future, we must endeavour to ensure not only that it is protected as an urban structure, but also that it is encouraged to support an active life.* In this connection it should be noted that, after their renewal, these historic 'vehicles of human activities' will be entirely capable of fulfilling a useful function. In this they differ from other 'objects' or 'containers of human activities' (means of transportation, tools, machines, etc.), whose preservation is both anachronistic and totally absurd once their technical capacity has been outstripped as a result of new developments (see 4115).

We have already discussed the role to be played by our living historic settlements in the future spatial cluster (see 222) and have explained why it is that these centres will be able to fulfil both traditional and new functions within the framework of future urban life. Despite the timeless[56] character of our historic urban centres as vehicles of specific functions (residential, cultural, tourist and artistic activities), their adaptation to the needs of their present populations and of the social classes who have to be attracted back into them (see 424) is a necessary condition of their survival.

This process of adaptation and regeneration calls on the one hand for the functional restructuring of both individual buildings and of the whole of the urban cluster, and on the other hand for a programme of urban renewal, which would entail sanitation projects and general restructuring.

By adapting our historic settlements in this way so that they are able to meet the present and future needs of their inhabitants and by impressing on the general public the aesthetic and cultural value of these centres, we can preserve their urban structure and also ensure the continuation of social life within their precincts.

---

55 A sad example of a historic town that has been deserted by its inhabitants is furnished by Mistra, the capital of the medieval Greek principality of Morea, which is now rapidly falling into a state of decay. This area was completely abandoned after the new town of Sparta was founded in the vicinity by King Otto I of Greece just over a hundred years ago.

56 'Timeless' in so far as they are able to adapt to the different needs and attitudes of different periods.

Up to about 1880, life in our historic settlements (small rural centres, homogeneous historic towns and historic urban sectors of major cities) had retained certain constant characteristics. These were:

a) *A high level of social symbiosis*, which was due to the high density levels prevailing in historic centres. Thus, although the living conditions were often extremely unhealthy, they also created a strong community feeling.

b) *Lively 'urban activity'*, which was reflected by the wealth of psychological experiences abounding on the shopping streets and public squares.

c)   *The total absence of motorized traffic.*

d)   *The almost total lack of large open areas* (parks, gardens, sports fields etc.) and of important social amenities (schools, nurseries, libraries, community centres).

e)   *The absence of big industrial concerns.*

Residential quarters, retail trading establishments, craft workshops, local administration offices and places of religious worship were all situated within the precincts of historic urban centres, where they contributed to the close-knit life of the community.

Then, about a hundred years ago, the character of urban life, which had remained constant for centuries, underwent a series of rapid changes which were brought about by the development of new social attitudes and technological methods[57]. The new demands made by urban populations created new needs and these led in turn to a restructuring of urban functions which altered the whole rhythm of life in historic urban centres. Urban interventions were the inevitable result of these new developments.

The changes which took place at this time and which were almost invariably confined to the overcrowded and overdeveloped areas of our historic centres had a disastrous effect on urban functions, for not only did they threaten the continued existence of individual historic buildings and disrupt the harmony of the urban cluster, they also seriously impaired the living conditions of the local population. With the building of large warehouses, the wholesale trade made its way into the historic urban sectors, where it soon took over extensive areas for the storage of goods. Various kinds of light industry were also set up, either in converted old buildings or in new buildings erected on old sites where they invariably constituted a discordant factor. Moreover, the tertiary sector (urban services), which had been developing throughout the whole of the nineteenth century, was definitely gravitating towards the historic centres.

In the closing years of the nineteenth century, the advent of motorized transport also posed a threat to our historic centres. Electric tramways, with their rails and overhead cables, and automobiles, which were then as now appearing in ever greater numbers in our towns and cities, hastened the decline of the traditional urban cluster and further impaired the living condition of its inhabitants.

In the course of the twentieth century, the traditional rhythm of life in our historic urban centres has been completely destroyed by the growth of motorized transport and has now been replaced by a new rhythm that is entirely characteristic of an industrial society. This development was accompanied by a determined – but futile – attempt to give our traditional urban centres new functions. The projects mounted to this end – which was pursued partly for sentimental reasons (popular attachment to the traditional townscape), partly for economic reasons (need to curtail civic expenditure) and partly from sheer lack of courage and imagination – have invariably disrupted the homogeneous structure of traditional urban space.

234 and 235
Two-storeyed private house at Clutter, England. The top picture shows the house completely defaced by advertisements designed to promote sales in the two shops. The bottom picture shows the same house after its restoration. The two shops are still there, but the advertisements have gone and the renovated shop windows have been integrated into the architecture of the façade.

57 These changes took place in homogeneous historic centres such as Florence and in the historic urban sectors of major European cities. The small rural historic centres were virtually untouched by this destructive wave of change. On the other hand, they suffered from constant demographic losses (see 424).

236

Mont-Saint-Michel in Brittany, which seemed predestined to be developed as a tourist centre. This monastery with its small medieval settlement forms a compact and independent entity which, because of its position on a peninsula, is not threatened in any way by urbanization. By introducing tourist centres and organizing cultural events within its precincts, this historic settlement could be given a new and socially useful function.

What was needed in view of the congestion of our urban centres and the uncontrollable rise in the population densities (in certain urban sectors of major cities, where the average height of the buildings is six storeys, these have reached the incredible figure of one thousand inhabitants per hectare) was not the unplanned introduction of new functions, which has simply led to the undesirable conversion of historic buildings, but a completely new urban infrastructure accompanied by adequate sanitation and urban restructuring (see 423). Because this was not done in the early years of the century, we now find ourselves faced with the urgent need to restructure and readapt the functions of our historic urban centres. Restructuring would involve the programming and reallotment of functions within the historic settlement while readaptation would involve a revision of the functions fulfilled by individual buildings. By effecting these changes we would be able to influence the character of urban life in our historic settlements so that it conformed to the structure and character of the historic centres themselves, which presupposes the existence of residential, cultural and artistic functions.

In the case of homogeneous historic towns and historic sectors of major cities, the most sensible way of programming urban activities would be to recreate as far as possible their traditional functions (residential areas, craft workshops, retail shops, cultural and artistic centres) and to abolish the disruptive functions which have developed in the course of the past one hundred years. Unfortunately, it would be extremely difficult to implement these measures in full, since there are so many vested interests concerned. None the less, if our historic urban centres are to be rehabilitated, they are an essential prerequisite. New functions which fit in with the unostentatious and peaceful character of historic centres can and must be introduced. This would call for the provision of artists' flats and studios, exhibition rooms and art galleries. Free access to historic centres must also be maintained (see 433) and certain amenities for tourists (hotels, furnished rooms, restaurants and pavilions) must also be provided within their precincts. But these tourist installations would have to be designed with great care to ensure that they did not disrupt the intimate character of the historic setting.

In the case of small rural historic centres, on the other hand, tourist installations should take precedence in the programming of new functions. These small provincial centres could be developed so as to attract tourists, for this would ensure the future of the urban community by providing work for the local inhabitants.

In view of the undeniable charm which historic centres hold for foreign visitors and – in the case of historic urban sectors of major cities – for the metropolitan population as well, it is only natural to consider whether these centres should not in fact be promoted as tourist attractions.

But the promotion of tourism, although desirable in itself, can easily get out of hand and might well become a disruptive element that would change the atmosphere of historic centres and gradually force the resident population to move out. The essentially temperate and peaceful character of urban life in a historic centre would be completely transformed if large numbers of restaurants, night clubs and gaming clubs were opened (see 4115), for these would introduce a misplaced note and probably create intolerable noise.

Consequently, the incorporation of tourist functions into a protected area must be carried out with the greatest possible care so as to ensure that a proper balance is maintained between the number of visitors and the number of permanent residents. As a general rule, the ratio for the whole of a given historic area should not exceed one visitor to every two residents during the tourist season, although in respect of certain particularly popular historic squares and at certain peak hours of day this figure might be allowed to rise to twenty visitors for every one resident.

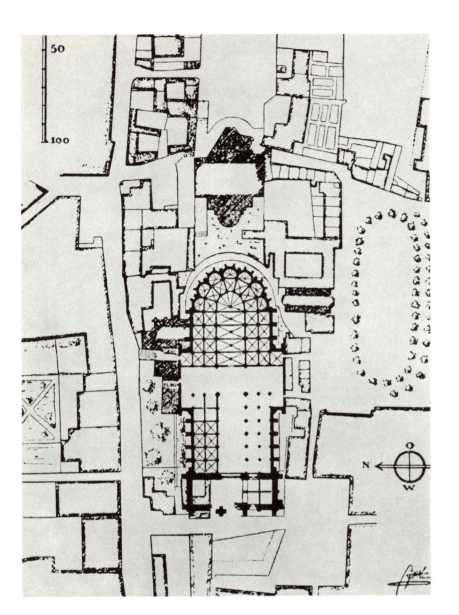

237
Plan of Cologne Cathedral in the early nineteenth century showing the squares in front and at the side of the building and the narrow streets surrounding the cathedral monastery. Approximate scale 1:2,000.

238
Present-day aerial photograph of Cologne Cathedral and its environs. The old buildings in the immediate vicinity of the cathedral were destroyed in the Second World War. But, long before this happened, the nineteenth-century planners had already defaced the townscape of this area and begun to destroy its architectural scale by erecting a gigantic railway station, a massive rail bridge across the Rhine and new residential blocks in the vicinity of the cathedral, where these symbols of the industrial age were sadly misplaced.

239

The principal façade of the administrative building of the Exarchate of the Holy Sepulchre in the Plaka, the old town of Athens. This building is in an excellent state of preservation and with its small monastery, church and courtyard is a focal point of present-day religious life. If it were to widen its activities, it could become a genuine cultural centre for the old town.

240

Façade of the new wing of a patrician house in the Plaka which the author restored, extended and converted into a restaurant in 1963. This house, which originally belonged to the Kalifronas family, is now run under the auspices of the Greek Minist ryof Tourism.

241 to 245

The Kalifronas family house before and after its conversion.

241   Plan of the ground floor after its conversion. The original two-storeyed wing is on the left, the new single-storeyed extension is on the right. Approximate scale 1:400.

242   Perspective sketch of the project.

243   Cross-section of the converted building. Approximate scale 1:400.

244   Cross-section of the original building. Approximate scale 1:500.

245   Plan of the ground floor of the original building prior to its restoration. Approximate scale 1:500.

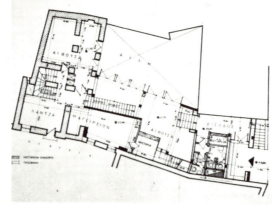

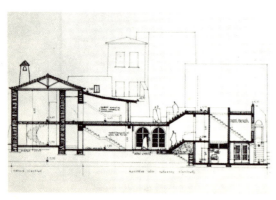

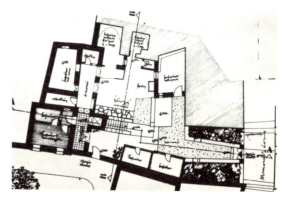

246
An undesirable case of vertical extension in the old town of Athens. This building now rises up in front of the Acropolis, where it is completely out of place.

247
Restored façade of the Kalifronas family house in the Plaka. The conversion of the building into a restaurant has not changed its architectural character.

248
A glaring example of the violation of a historic centre as a result of tourist 'promotion'. This completely shapeless hut, which was built from an assortment of different materials in the heart of the Plaka, is a pseudo-picturesque *mise-en-scène*.

249
A rather unfortunate conversion in the old town of Zürich. The incorporation of a boutique into this traditional house has disrupted the architectural organization of the façade due to the arbitrary disposition of the new apertures.

250
A more successful conversion in the old town of Zürich. Here the whole of the ground floor of a private house has been turned into a restaurant. The architect has made an honest and intelligent attempt to integrate the form of the new large windows into the general design of the façade (by the use of partitions and suitable materials).

251

Photograph of the interior of an Italian monastery after its conversion into a museum.

In the readaptation of individual buildings, an equally necessary process, numerous interesting and, in some cases, quite daring experiments have been carried out by individual property owners and architects. Recently a thesis has been advanced which has met with fairly wide support to the effect that the value of a building lies mainly in its architectural and historic form. This form, it is argued, must be preserved at all costs when a building is renovated and readapted, whereas its functions may be completely transformed if this contributes to the revival of the building. The champions of this new theory, who regard themselves as 'modern', say that they do not intend to be inhibited by aesthetic or artistic considerations. But it is rather surprising to find that after the dictatorial protagonists of 'modern architecture' had announced in the first half of the 1920s their universal aesthetic principle of functionalism (according to which architectural form is simply a derivative of architectural function), their spiritual heirs are now maintaining the precise opposite, namely that 'the form of a building is all-important and that we are at liberty to change its functions to suit our own ideas'.

In point of fact, neither of these extremes is valid. The truth lies somewhere between the two. It goes without saying that, by means of conversions, alterations and extensions, we are able to adapt any building to fulfil almost any new and alien function. But, before doing so, we should ask ourselves:

a) How much of its original architectural form and substance will survive.

b) Whether by subjecting its traditional function – which is accepted and respected by the general public – to ruthless alterations, we are not betraying the essence of the building.

In the case of private dwellings, sheds, warehouses or cellars, i.e. buildings or rooms of no great historical or architectural significance, we would endorse their functional readaptation. Such buildings could profitably be converted into restaurants, clubs, workshops and retail businesses.

Moreover, old deserted monasteries of little historical or artistic value and old hospitals, asylums and prisons, which are no longer required to fulfil their original function, are suitable subjects for conversion into hotels or cultural and artistic centres.

But to try to change the function of a building, which is well known to the general public and which also possesses artistic value, is a very difficult and hazardous undertaking which seldom succeeds, save when the new function is closely allied to the original one (for example, the conversion of a Renaissance palace into an administrative centre or of an eighteenth-century private house into an ambassadorial residence). But, where there is no such continuous link, functional readaptation calls for major alterations both in the layout of the rooms and in the internal and external form of the building[58]. It also offends public opinion, which rightly regards such conversions as arbitrary and unacceptable.

---

58 A typical example of unsuccessful readaptation is the use of historical palaces as museums, in which art objects are displayed that are often completely alien to the character of both the building and the country in which it is situated. The Louvre in Paris was misused in this way for decades: although the lighting in its rooms and the rooms themselves were quite unable to answer the needs of a museum, they were used for the display of oriental and Greek antiquities. Today the situation has been reversed: by readapting the rooms in the Louvre to meet the requirements of a museum, the historical organization of the interior of this famous palace has been subjected to arbitrary changes.

252 and 253

Two extremely daring but highly regrettable replicas of Flemish Rococo houses, which now accommodate commercial functions at ground floor level. These two façades are, of course, stylistically absurd. Although they are meant to look traditional, they do not have a traditional foundation and so appear to be suspended above the contemporary shop windows on the ground floor.

254
Inner courtyard of a tenement block in the Berlin district of Wedding. These blocks, which are now about one hundred years old, are proper subjects for urban renovation.

255
Colony of huts at the foot of the Acropolis. These huts, which are known as the Anafiotika and were built as temporary homes between 1836 and 1840 by masons from the Aegean island of Anafi when they came to Athens to work on the Royal Palace, are still in use today.

256
A picture of despair. The centre of Augsburg after the air raids in 1945.

It is clear, therefore, that the conversion and functional readaptation of important historic buildings should be undertaken only with the greatest care and that before any such project is launched, the architect concerned should consider to what extent his proposed functional alterations will be accepted by the public. In view of the significant loss of architectural form which must be incurred as a result of these conversions, we should approach all such undertakings with great reserve.

The theory that functional readaptation is always legitimate provided it 'brings the building back to life' is likely to lead to the deformation of buildings, which in our opinion is a far more serious threat than that posed by progressive but dignified decay.

423   *Urban Renovation. The Sanitation of Buildings and the Restructuring of the Urban Cluster*

Functional readaptation is the first technique employed in order to bring our historic centres into line with the needs of present and future urban life. The second is the sanitation of buildings, which is coupled with the general or partial restructuring of the urban cluster.

During the past twenty years these methods have been used as a first step in the renovation of the urban clusters in our major cities. As a result it has been possible to evolve a new methodology for improving living conditions in old and unhygienic urban sectors.

This system was first employed in different kinds of urban areas which had nothing to do with historic centres. These were:

a) Urban sectors of major cities, whose old-fashioned buildings (mostly over one hundred years old) had never been fitted with proper sanitary installations and whose general fabric was also in a lamentable condition, due partly to the indifference of the landlords (which was due in turn to the freezing of rents during the two world wars) and partly to overcrowding.

257 and 258

Houses at Brecon, Wales, before (257) and after (258) renovation. When modern sanitation was installed in these houses, the old plaster was also removed from their façades, revealing the original timber framework.

b) Rural settlements, which had never received a modern infrastructure and whose buildings had seldom, if ever, been renovated.

c) Urban sectors which had been damaged or destroyed during the two world wars.

d) The unhygienic shanty towns, which are still to be found on the outskirts of many major cities and whose continued existence is a social disgrace.

Although an analysis of the intervention techniques employed on these projects would exceed the limits set for this book, it should be mentioned that the sanitation and urban restructuring which needs to be carried out in our historic centres is a minimal task compared with the work that has yet to be done if all the old-fashioned urban clusters are to be brought up to date.

It should also be mentioned that these methods often have to be applied in a way that would be far too radical for the valuable buildings in our historic centres.

The sanitation of individual buildings, which is intended to provide them with modern facilities and to reorganize their interior design along functional lines, consists of:

a) Regular maintenance and, where necessary, strengthening of the structural members of the building (see 4122).

b) Conversions rendered necessary by the introduction of sanitary installations and water supplies and, in certain cases, central heating and electricity.

c) Structural adaptation (enlarging windows and doors, lowering ceilings, changing the dimensions of rooms) to bring the various rooms into line with contemporary design and functions.

d) Partial or total transformation of interior and exterior decorative motifs and paintwork to make them look more 'modern'.

Most of the accompanying buildings in historic centres (see 4121) are residential and, as far as they are concerned, the first two processes listed above (maintenance work and sanitary installations) should certainly be carried out in order to provide the occupants with a modicum of comfort.

The other processes, i.e. structural adaptations, are also desirable in so far as they would bring the form, dimensions and character of these buildings more into line with present-day standards. These could also be carried out in all cases provided the façades of the buildings were not altered.

Since these buildings are not architectural monuments, it is entirely reasonable that part of the original form and substance of their interiors should be sacrificed in order to bring them up to date. This concession is not at variance with the general aims of rehabilitation, which is not intended to preserve the interior design of 'accompanying buildings' but merely to ensure that their façades – which make an important contribution to the townscape of historic centres – remain unchanged.

In the case of architectural monuments, on the other hand, interventions should be confined wherever possible to maintenance and conservation work. Sanitary installations, interior conversions and other major alterations are permissible only if they do not impair the morphological integrity and historical value of the architecture. Needless to say, such work should be carried out with extreme care.

Alterations to decorative motifs and to paintwork, which are often made on buildings of no historic or artistic value in order to 'modernize' them, are not permissible either on architectural monuments or on their accompanying buildings. Alterations to the façade of any building in a protected area undermine the authenticity of its historic architecture.

Such alterations are, of course, not only permissible but highly desirable where buildings are 'stripped' (see 4125) in order to recreate a particular phase in their historical development.

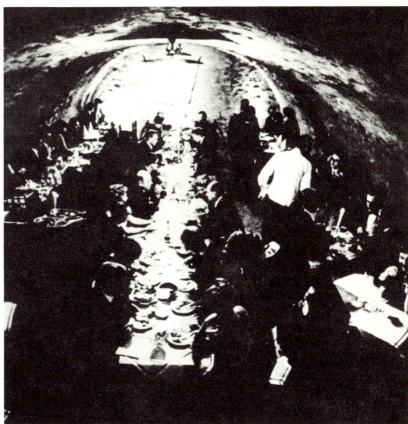

In conclusion we may say that although the sanitation of buildings in historic centres is desirable in itself, this does not mean that it may be systematically pursued as it is in decaying urban centres which possess no historic value. On the contrary, sanitation works in historic centres should be permitted on condition that they do not impair the traditional townscape.

This condition is very important and needs to be understood and accepted by those town-planners who wrongly believe that the general sanitation and restructuring of urban clusters is identical with the rehabilitation of historic centres (for a definition of these concepts see Summary p. 184).

The restructuring of urban clusters which, like the sanitation and modernization of individual buildings, is designed to improve urban living conditions, is often an extremely radical undertaking. It involves the following processes:

a) The demolition of neighbouring buildings and outbuildings to give more air and sunshine.

b) The demolition of whole blocks of dwellings which are in the final stages of disintegration. The sites freed in this way are then available for new blocks (which can be freely aligned, i.e. need not follow the line of the roads) and for urgently needed urban installations (parking spaces, garages, gardens, sports fields, etc.).

c) Radical alterations to the road networks, involving extensions and link roads. It is hoped that the creation of urban clearways and of new access roads to city centres, coupled with the provision of additional parking space, will bring our road networks up to date and enable them to cope with the greatly increased demands of the motor car age. A parallel development is devoted to the creation or recreation of pedestrian walks and areas.

259
An attic room of a traditional house in Gamla Stan, the old sector of Stockholm. In fitting out the room to contemporary standards, the architect has made a special feature of the roof timbers.

260
'Fratis Kellaere', a smart restaurant in the vaulted cellar of a converted house in the old sector of Stockholm.

261 and 262

During the Second World War, the timber roofs and ceilings of many of the old houses in Potsdam were burnt out. The diagrams on the left show a cross-section of such a house prior to renovation, together with plans of the ground and first floors. The diagrams on the right show the fitting-out of the area immediately below the timber roof and the re-arrangement of the rooms.

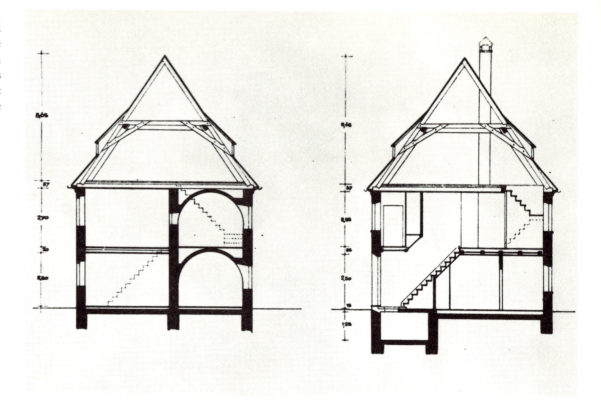

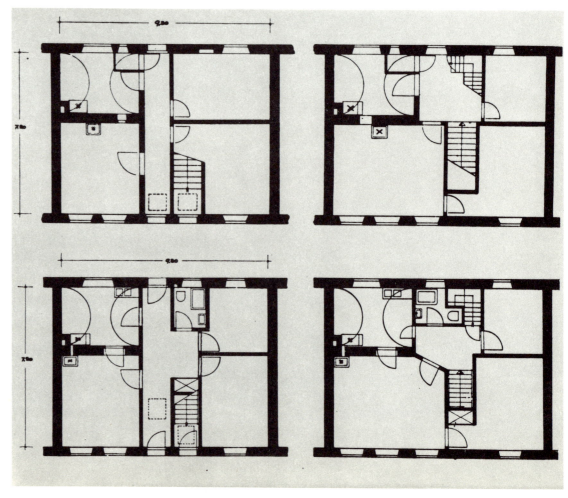

263 and 264
Apartment house in Berlin before (263) and after (264) the renovation of its façades. Fifteen years ago, it was widely believed that a building could be renovated or 'modernized' simply by removing its original stucco. In point of fact, all that this achieved was the complete deformation of its architectural character.

265
The radical restructuring of the urban centre of Worcester, England, is an object lesson in how not to intervene in a historic area. By completely ignoring the architectural scale of the original composition, the planners have destroyed its structural unity.

266 and 267

The urban sector of Maubert in Paris is an example of how interventions in historic centres should be carried out. The restructuring of this district has been extremely well planned. The left diagram (266) shows the present condition of the buildings. The right diagram (267) shows the new scheme. It is proposed to reduce the density of urbanization and to create passageways and gardens by 'gutting' the inner courtyards of certain blocks. Approximate scale 1:4,000.

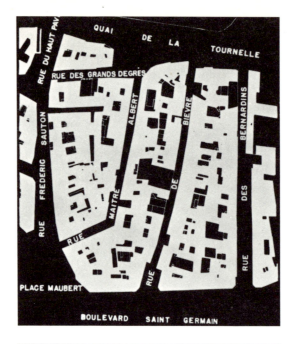
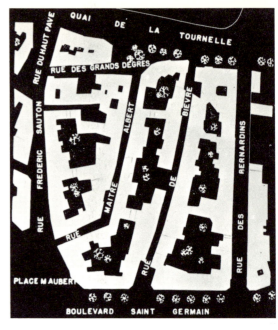

268 and 269

The Hansaviertel in Berlin. This district was almost totally destroyed during the Second World War and has now been radically restructured. As a result it has acquired a completely new townscape. This kind of restructuring would not be permissible in a protected area.

268 The Hansaviertel in the 1930s. The buildings were set out in compact blocks with their principal façades parallel to the road.

269 The Hansaviertel after its reconstruction, which was launched at the time of the *Interbau* exhibition in 1958. The district now contains freestanding buildings arranged asymmetrically.

270 to 275 (p. 153)

The urban sector of Wedding in Berlin. Here a radical restructuring programme was launched which soon developed into a reconstruction project. All the derelict nineteenth-century buildings in the district were demolished in stages. This procedure, which would have been unacceptable in a historic centre, was permissible in Wedding where the original buildings had little or no aesthetic value.

In the case of historic settlements, restructuring is very much more difficult because of the need to protect not only the historic buildings, but the townscape and the specific structure of the historic nucleus as well.

Consequently, urban renewal in historic centres should be largely restricted to the sanitation of individual buildings. Restructuring of the urban cluster, when it takes place at all, should always be discreet; few buildings should be demolished and few new roads built

For all practical purposes, therefore, we can dismiss the idea of a restructuring of the urban cluster as far as historic urban centres are concerned, for any attempt to implement such a project would cause irreparable harm both to the authentic substance of the buildings and to the urban cluster itself. Rather we should think in terms of the 'urban rehabilitation' of our historic centres, which would involve their urban regeneration (using the various techniques discussed in sections 41 und 42) and their integration into neighbouring urban areas. The problems posed by this process of rehabilitation will be dealt with in the final section of this study.

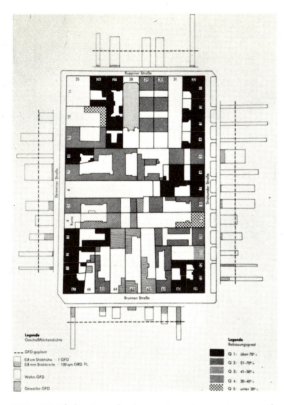

270 Initial density of urbanization. Approximate scale 1:3,500.

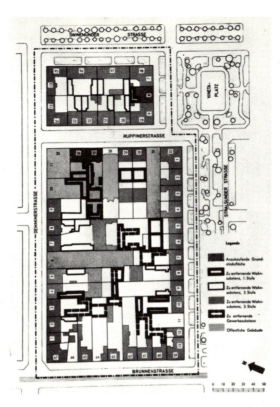

271 Evaluation of sanitation requirements. First phases (demolition). Approximate scale 1:3,500.

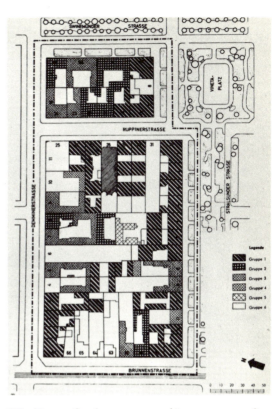

272 Types of settlement arranged in groups according to quality. Approximate scale 1:3,500.

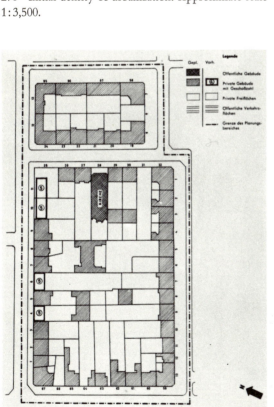

273 Restructuring: end of the first phase of reconstruction (conversions and new buildings). Approximate scale 1:3,500.

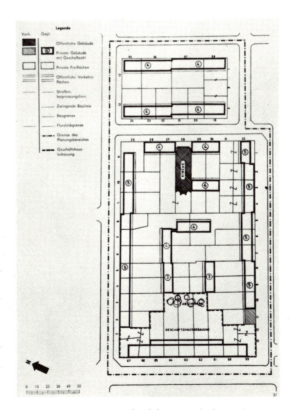

274 Restructuring: end of the second phase of reconstruction (replacement of all buildings). Approximate scale 1:3,500.

275 Aerial view of Berlin-Wedding.

Over the past one hundred years, important changes of function have taken place within our historic centres and these in their turn have brought about important demographic changes.

This development has been least marked in the small rural and provincial historic centres, where there has been a gradual but constant drop in the population due almost entirely to the call of the city which prompted these country people to leave their traditional homes and become city dwellers. It is highly desirable, if not essential, that new functions – especially tourist functions – should be introduced into these historic settlements so as to create work for the local population and halt this exodus.

The demographic changes which have taken place in the inner areas of homogeneous historic towns and – more especially – in the historic urban sectors of major cities have been more serious. Here the principal factor is the advanced age and poverty of the remaining inhabitants.

The people who stay on in an old residential district that has been degraded by the installation of warehouses, wholesale businesses, and small factories, are almost invariably the older members of the community. Whether for sentimental reasons or because they feel unable to adapt to new conditions, the older inhabitants tend to stay on, while the young people prefer to live in new districts – even if this means moving to one of the outer suburbs – provided they are able to obtain technical and social amenities and are assured of better living conditions.

The degradation of the inhabitants of historic urban nuclei is a phenomenon that is as widespread as it is lamentable. The relatively wealthy middle classes are gradually leaving these old residential districts with the result that they fall into an ever greater decline. Rents are being pushed down and tradesmen, manufacturers and wholesalers are taking over the apartments vacated by the middle classes. The oldest and most dilapidated buildings are being reduced to slums[59].

It is because conditions are so bad that an attempt is now being made to provide acceptable living conditions by carrying out sanitation projects and by rehabilitating these inner areas of our historic urban centres. In this way it will be possible to prevent the demographic demise of these urban sectors. We know from experience, *once these measures are taken, the demographic trend is immediately reversed. After the disruptive and incongruous functions have been removed from a historic urban centre and the houses have been renovated, the rehabilitated townscape exercises an instantaneous effect with the result that the original inhabitants are completely captivated and only too eager to return. A wave of recolonization takes place so that rehabilitated historic centres, having regained their original function, are also regaining their original population.*

It is often argued that however beneficial the effects of sanitation and urban rehabilitation may be for the revival of social life in protected areas, these measures inevitably raise land and house prices. The conclusion is then drawn that urban sectors of this kind are being developed into residential districts in which only the privileged classes will be able to live. This is perfectly true. But, far from harming the historic centres, this state of affairs will actually ensure them a long lease of life.

59　Typical examples of this regrettable development are to be found in the historic urban sector of the Marais in Paris which – because of its proximity to Les Halles (central food market) – was gradually transformed into a centre for wholesale business, craft workshops and light industry despite the fact that this sector still contained magnificent buildings dating from the seventeenth century, which were allowed to languish in filth and squalor. Further examples are furnished by the eighteenth-century urban sectors of Psiri and Monastiraki in Athens, which died a lingering death due to demographic changes. The Plaka, the old town of Athens, is a special case, which should also be briefly mentioned. Over the past ten years the Greek Archaeological Service gradually bought forty properties in the heart of the Plaka with a view to carrying out archaeological investigations at some future date. Meanwhile, however, farm workers and their families, who left the countryside to seek work in the capital, have moved into the empty houses as squatters. They now lead a nomadic existence rent-free in the heart of the city, where they are contributing to the degradation of the social life of the Plaka.

The changes which are bound to take place in the social make-up of protected areas after their regeneration (due to people moving in and out) will be based on our western principle of supply and demand and will, therefore, not involve any sudden demographic shifts.

But there is still one last question which has to be answered in this connection, namely to what extent will regenerated historic urban centres be able to play a dual role? On the one hand they will have to fulfil a subdued function as residential and cultural centres, while on the other hand they will have to attract tourists, who will inevitably introduce an element of noise.

The well-preserved and centrally situated historic urban sector of Saint-Germain in Paris and the old towns of Zürich and Geneva, which are constantly visited by large numbers of foreign visitors but have still not lost their intimate character, are an encouraging sign. Judging by them, it would seem that these two diametrically opposed functions can certainly be combined.

425　*Socio-psychological Problems and Questions of Aesthetic Education*

The most crucial factor in the whole field of preservation is undoubtedly the attitude of the inhabitants of historic centres (and of neighbouring urban sectors) to their protection and revival. In view of the cost and complexity of interventions in historic urban clusters, it is also certain that all such projects must be organized and executed by national, regional or local authorities with the help of specialist firms. But the success of such operations depends to a very large extent on the sympathetic understanding and the moral and practical support of the population.

As has already been indicated (see 131), the inhabitants of different historic towns and urban sectors often adopt widely divergent attitudes to the question of preservation. This is due partly to variations in local conditions, and partly to variations *in the cultural and intellectual levels of different urban groups.*

In many cases the inhabitants of historic centres are only too eager to move out of their traditional houses or, alternatively, to have them demolished and replaced by new buildings that clash with the urban setting. This tendency is due partly to a lack of aesthetic sensitivity and partly to an understandable desire for modern living conditions. On the other hand, we also find – usually amongst the inhabitants of small rural historic centres – the reverse tendency where people are deeply attached to their old familiar environment and to their traditional way of life. As for the regenerated historic residential areas, their preservation is assured by the general affluence of their inhabitants.

It is regrettable, however, that in our liberal western world the sacred aura which surrounds private property should have proved such an obstacle to the administrative protection of historic centres. Whereas the legal protection of classified architectural monuments (belonging to the state, to local communities, to the church or to private individuals) has long been accepted without demur, the public is not so easily convinced of the need to extend this protection to whole historic urban areas as well. Moreover, all attempts to restrict the absolute right of house or land-owners in respect of their property (by requiring them to obtain special permits for demolition work or alterations to buildings in traditional areas) have met with resistance and not infrequently provoked considerable hostility. This is one of the reasons why so little progress has been made in Europe in the formulation of effective legislation for the protection of historic settlements and urban centres. The only country to have made real headway in this respect is France, where the 'Malraux law' (Law No. 62.902 of 4 August 1961) provides for the legal protection of historic settlements.

The present attitudes of the inhabitants of historic settlements, although understandable in themselves, are in many cases mutually exclusive. Some want to live in a new district with modern sanitation, others insist on their absolute right to dispose of their property as they see fit, while others yet are genuinely attached to their traditional way of life. What is needed is a rational education programme, geared to the needs of the various social classes, which would provide a basic training in aesthetic appreciation, thus enabling these people to adopt a sympathetic and informed approach to the rehabilitation of their historic centres.

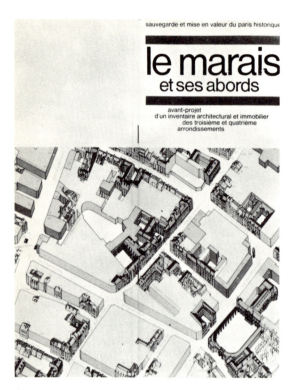

sauvegarde et mise en valeur du paris historique

le marais
et ses abords

avant-projet
d'un inventaire architectural et immobilier
des troisième et quatrième
arrondissements

276
Axonometric aerial view of the urban sector of the
Marais in Paris. (Published by the *Association de la sauve-
garde et de la mise en valeur du Paris historique*.)

Many attempts have already been made to tackle this problem both by societies and by groups of private individuals. These workers have also undertaken conservation and restoration projects on their own account and, although their operations in these two spheres have been necessarily restricted by lack of adequate funds, they have undoubtedly made a major contribution to the task of creating a favourable climate of opinion with regard to rehabilitation[60]. By organizing festivals, exhibitions, lectures and a wide variety of performances, these societies and groups try to arouse public interest in the fate of the historic centres. *This publicity makes people aware of their collective responsibility and, for the time being at least, that is enough to ensure that our historic centres are spared the worst ravages of private intervention.*

As far as the future is concerned, it is absolutely essential that we should gain the confidence of the public. We must create a climate of opinion in which it will be difficult not to approve of rehabilitation. Public discussions and demonstrations must be organized and wherever possible, these should be held in urban sectors which have been earmarked for rehabilitation. Thus the inhabitants would not only be informed about the artistic value of historic urban areas, they would also be given tangible proof that provided they are rehabilitated, such areas will be both viable and attractive. By proceeding in this way, we could create a new mental rapport between the individual and his architectural environment.

60 The *Association des Amis du Paris Historique* is one such group. In 1962 it organized its first Marais festival, which convinced the French authorities of the need to protect and revive this beautiful urban sector.

431   *Demolition in Protected Areas*

Apart from acts of war and acts of God, there are two principal reasons why individual buildings or groups of buildings are demolished in historic settlements. These are:

a) *The sporadic and uncoordinated erection of new buildings on the site of historic buildings. This is an undesirable development carried out by private individuals which should be opposed whenever possible.*

b) *The planned restructuring of historic urban clusters (see 423) and the rehabilitation of their townscapes. This is a necessary and desirable development.*

The unplanned demolition of houses came about as a result of land speculation. Certain private landlords with property in historic settlements have had traditional buildings demolished in order to use their sites for new buildings which bring a higher return on capital. Moreover, when a building is demolished in this haphazard fashion, the site is often left empty for some time, thus disrupting the continuity of the townscape. But, even if the new building is erected at once, we are still faced with the difficult aesthetic problems posed by modern structures in a historic area (see 4122).

It is highly desirable, therefore, that private individuals should be prevailed upon to desist from the demolition of property in historic urban areas. Although the only way of acquiring complete control over private property is by taking out a compulsory purchase order which does, of course, involve compensation, this would not really be to the purpose, since it would inevitably lead to the demise of the historic centre concerned. Fortunately, such extreme measures are not indispensable, for there are other ways of tackling this problem. An attempt should certainly be made to win over these private owners and enlist their support. A propaganda campaign designed to enlighten the inhabitants of historic centres and enable them to appreciate the aesthetic quality of their environment would undoubtedly activate public opinion and so bring pressure to which many might well respond. However, to ensure complete success it is imperative that special building regulations should be introduced for all protected areas. By stipulating that the total useful capacity of any new buildings should not exceed that of the buildings which they replace, the authorities could effectively solve this problem, for then there would be no financial incentive to demolish an old building in order to build a new one.

The planned demolition of buildings in historic urban centres is undertaken primarily in order to remove derelict unaesthetic or stylistically incongruous buildings. Where derelict buildings are concerned, the authorities in most European countries are legally entitled to proceed with demolition work if the buildings are dangerous. 'Unaesthetic' buildings can also be disposed of in this way, but only after a compulsory purchase order has been taken out. The sites freed as a result of planned demolition can, of course, be used either for parks and gardens or for new housing developments. But, although demolition projects carried out with a view to restructuring historic urban centres are often highly desirable, this method of urban renewal can never be used on the same scale as in urban areas of no historical or cultural value (see 423) since most traditional buildings – even those in a bad state of repair – ought to be conserved rather than demolished.

277

The market place in Henley-on-Thames, England. The demolition of just one building and its replacement by a single-storeyed shop was enough to destroy the streetscape, which was determined by the rhythmical repetition of façades of equal height running parallel to the road.

278

In Marlborough, England, on the other hand, the architect who designed the new Woolworth building created a roofline which fitted in with the general streetscape. He also broke down his wide façade into three equal sections, which brought it more into line with the traditional façades on either side and ensured its integration into the urban image.

In general, interventions for purposes of urban restructuring fall into two classes:

   a) *The first of these involves a special system of 'gutting', i.e. removing the inner core of residential blocks, which has been developed of recent years and is now being used in various countries.* This calls for the removal of all annexes such as studios and storerooms which were built – often as a temporary expedient – in the inner courtyards of these blocks, and all shops or extensions superimposed on to their façades, which detract from the authenticity of the original building. Thus, without even touching the original substance of the buildings, we are able to increase the amount of light and sun which they receive, raise the value of the apartments overlooking the courtyards, rehabilitate the façades and greatly reduce the density of urbanization in the area.

As a result of these measures, the courtyards to the rear of the apartment-blocks would be left completely bare and could be restructured to provide open areas with benches and lawns, play centres for young children etc. Incidentally, it would not be necessary for the community to appropriate these courtyards since they could be rented from their present owners. Open or enclosed passageways, which might possibly involve compulsory purchase, should also be built to link the open areas in the courtyards with the traditional road network.

This type of intervention makes a major contribution to the rehabilitation of the townscape for it increases the integration of buildings and open areas in the urban cluster, creates new façades in the inner courtyards, and also produces additional perspective views, which provide new visual experiences. With their free alignment, the 'rediscovered' courtyard façades constitute a novel and intriguing visual element.

In certain cases, where a building has to fulfil only a single function (e.g. as a cultural or tourist centre) and where its overall dimensions make it a suitable choice, new buildings can be designed for historic centres. The Studentenhaus [Students' House] in Regensburg (see fig. 220) is a case in point. Buildings of this kind – which must fit in with the architectural style of the urban cluster – are usually erected in the midst of a residential quarter, where they form a harmonic contrast to the historic buildings around them.

   b) *The second class of intervention carried out in the restructuring of urban centres involves the demolition of buildings which detract from the original urban design or which block the view of important architectural monuments or groups of buildings.*

158

279 and 280
Conversion of four residential blocks in the 'operational sector' of the urban district of the Marais in Paris. Many buildings within the blocks have been demolished and replaced by new structures and the inner courtyards have been landscaped. This project, which has been very carefully planned, forms part of a systematic and extremely sensitive restructuring programme for the whole historic sector of the Marais.

279 Model of one of the four residential blocks (top right on the plan. See fig. 282) after the removal of the derelict interiors and the landscaping of the inner courtyards. The new replacement structures are based on contemporary principles of design but are on the same scale as the original structures.

280 Another view of the same model.

281
Model of all four residential blocks after their restructuring. A large number of new structures – e.g. underground garages and small gardens – will be installed in the inner courtyards. The demolition of the interiors of these blocks will not affect the basic character of the townscape since most of the new structures will be internal. They will also be built on the same scale as the original buildings, and will therefore be closely integrated into the historic townscape.

282
Plan of the 'operational sector'. Black areas: rehabilitated historic buildings. Grey areas: new structures. Dotted areas: new gardens. White areas: flagstoned courts. Approximate scale 1:3,000.

We frequently find historic squares which have been hemmed in by buildings of a later date and consequently have lost their original character. Even when these later buildings possess certain architectural value, it would probably be better to eliminate them in order to re-establish the original unity and clarity of the historic urban area.

Buildings of little or no architectural value which are situated in historic centres may also be demolished in order to create new perspective views of the general townscape[61] or of individual monuments, in order to extend the squares on which such monuments are situated (the medieval cathedrals are an exception to this general rule), and in order to create new views from points within the urban cluster of the principal features of the surrounding landscape.

---

61 For the urban rehabilitation of the Plaka (see fig. 297), the author suggested that a row of houses on the Hadrian road should be demolished. This would make it possible to create a green belt area and also to excavate the foundations of the fourth-century Heruli city wall, which would then form a symbolic boundary between the old and new town. The green belt would also allow observers passing along the Hadrian road unimpeded views of the Plaka and the Acropolis.

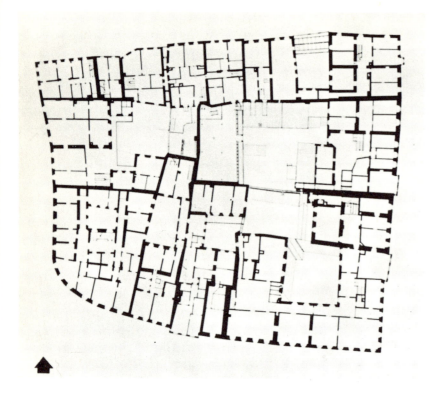

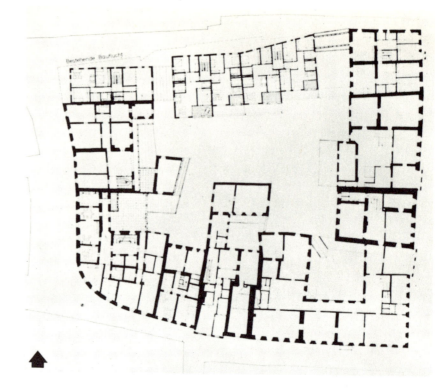

283 to 285c
Analysis of the existing structures and of the restructuring of a residential block, which formed part of the detailed study for the urban rehabilitation of Regensburg carried out by a group of architects headed by Professor Werner Hebebrand between 1965 and 1967.

283   Plan of the existing first floor. Approximate scale 1:800.

284   Plan of the proposed restructuring of the first floor (involving the 'gutting' and redesigning of the interior and the construction of a new northern façade). Approximate scale 1:800.

285a   Present exploitation of the ground, first and second floors and of the whole building (cross-section). Approximate scale 1:3,000.

285c   Planned exploitation of the ground, first, second and third floors. Approximate scale 1:3,000.

285b   Number of storeys, condition, age and cultural value of the block. Approximate scale 1:3,000.

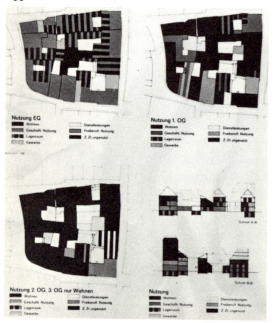

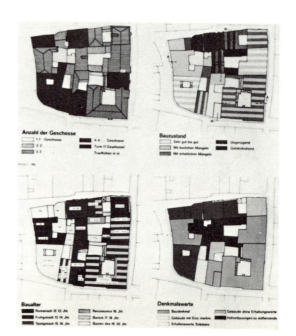

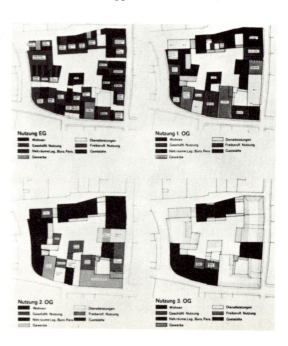

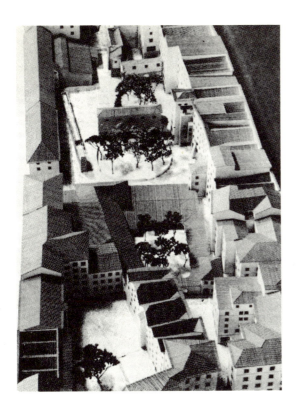

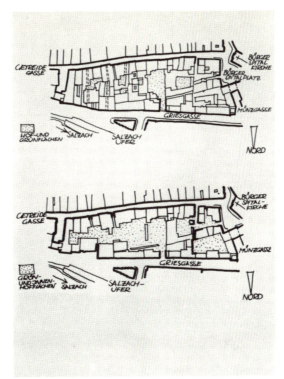

286 and 287
A restructuring project: residential block in the old town of Salzburg.

286 Aerial view of the project.

287 Top: original condition. Bottom: condition after the 'gutting' and restructuring of the interior.

288 to 290
Two sectors in the protected area of Bourges (France) which are now being rehabilitated.

290 Plan of the restructured sectors.

288 Plan of the protected area. The sectors in question are at the top left and are enclosed by thick black lines.

289 Working plan. Black areas: preserved and rehabilitated buildings. Dotted areas: derelict and unimportant buildings earmarked for demolition. Grey areas: new buildings. White areas: courts and gardens. Approximate scale 1:2,500.

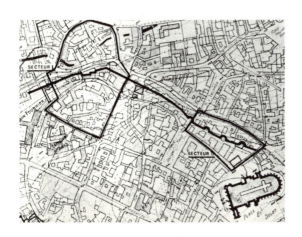

When we speak of introducing a modern technical infrastructure into a historic centre, we mean the installation of public electricity, fresh water and central heating networks together with the provision of sewers connected with the main system in the newer urban sectors of the town. In the majority of historic settlements, such networks are either inadequate or non-existent. The introduction of suitable technical infrastructures is regarded as an essential part of urban rehabilitation since, without it, our historic centres could not comply with present-day hygienic requirements and their survival would, therefore, be prejudiced on these grounds. The provision of new road networks and new transportation systems also comes under the heading of modern technical infrastructures, but these will be dealt with in the next section (see 433).

As far as the supply of fresh water is concerned, most historic centres still have old or, in some cases, antique installations (e.g. Roman aquaducts), which carry drinking water from artesian wells, from large storage cisterns or from the water mains laid in neighbouring new sectors to different points within the urban cluster, where it can be drawn off from fountains or pumps[62]. *As far as possible these picturesque traditional installations should be kept in working order since they often constitute an important visual element in the general layout of an urban square or street.* At the same time, however, it is essential that every single house within a historic centre should be provided with fresh water from the tap, for this is an indispensable condition of urban sanitation.

---

62 In certain central districts of Berlin, which were laid out in the nineteenth century but which have long since been provided with tap water, large hand-operated pumps made of cast-iron are still to be found on the pavements.

292a Section a–a of area after its rehabilitation. Approximate scale 1:1,000.

292b Section b–b of area after its rehabilitation. Approximate scale 1:1,000.

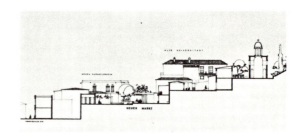

291
Restructuring of a residential block. Part of the proposed rehabilitation of the town of Bietigheim in Württemberg. Five alternative schemes were submitted for this block. These ranged from conservation and conversion projects to a plan for its total reconstruction. Approximate scale 1:3,000.

292a to 293d
The area around the 'Old University' (which was actually the house of the architect Kleanthis) in the Plaka in Athens.

292c Model of the area in its present condition. Top right the 'Old University', top left the Church of the Blessed Virgin 'Chrysocastriotissa'. The average width of the roads and paths is four metres. Many of them are stepped. The residential blocks are very small (average area of site 400 square metres) and the buildings, which are either one or two-storeyed, have highly irregular forms. This, coupled with the small architectural scale of the area, produces an extremely intimate townscape.

292d Scheme proposed by the author for the rehabilitation of this area based on the principle of harmonic contrast, and involving the planned demolition of specific buildings and their replacement by new structures geared to the architectural scale of the area. The house

of Kleanthis would be preserved as an architectural monument of recent times, and used as the main building of a cultural centre. This centre would embrace the whole of the Kleanthis complex and two neighbouring blocks. Small footbridges would give direct access above street level between these three sections. The vaulted rooms on the ground floor of the 'Old University' would form an extension to the basement restaurant of the cultural centre. The first floor of the centre would house the administrative offices, conference rooms and a small library, while the second floor would be fitted out as a local history museum containing exhibits from the old town of Athens.
The upper part of the neighbouring block (in the centre and on the left of the house of Kleanthis), parts of which

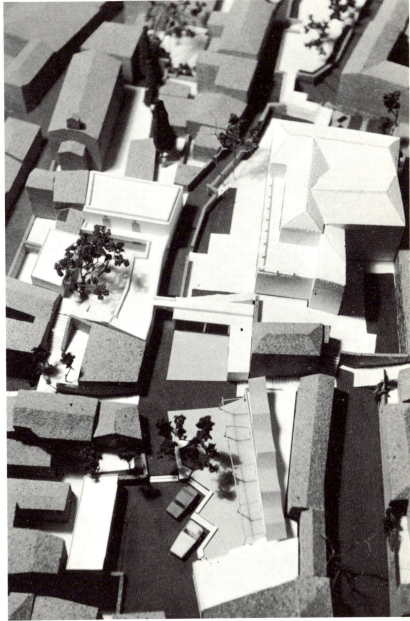

are now derelict, would be restructured so as to accommodate a caretaker's flat and a small artists' centre with four studios and an exhibition hall. All the new structures would be in a contemporary style but their layout, height and scale would all take account of the historic urban setting. Part of the site occupied by the third block (at the top of the model in front of the 'Old University') would be laid out as a garden with a small covered café where the artists could relax. It is also proposed to install a small market which would open on to a small public square. This is one of three different sites suggested for the creation of neighbourhood markets in an as yet unpublished study by the author, dealing with the protection of the whole of the Plaka. This square would be the terminal point for a branch road (running into the Plaka

from the main access road) which would be reserved for the delivery and collection of goods. All other roads would be reserved for pedestrians (see also commentary on fig. 331). The site for the square and for the new shops does not exist at present and would have to be created by the demolition of three houses which possess no architectural value. In order to make the greatest possible use of this site, which has a natural slope, the shops would be built at basement level beneath the courtyards of existing houses. At the far end of the market place a new two-storeyed shop would be erected, which would form the boundary for that side of the square and would also conceal the side wall of a two-storeyed Neo-classical house that is to be preserved. No other alterations to the existing road network are envisaged under this scheme.

The narrow, stepped alleys would all be preserved. The medieval city wall, which runs down the slope behind the 'Old University' and is now completely hidden from view, would be exposed and trees would be planted behind it to heighten its visual impact.

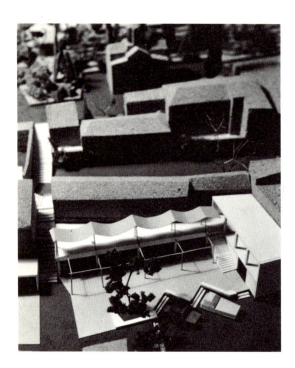

293d Detail of the model. The new market place.

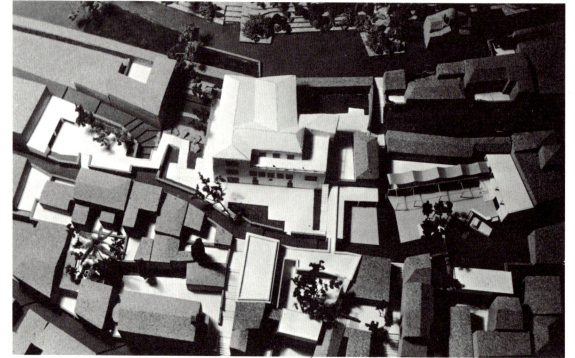

293a Klepsyda Street. View of the street front showing the cultural centre on the left and the small market place on the right. Approximate scale 1:800.

293b Model of the whole area after its conversion.

293c Ground plan of the proposed scheme. Approximate scale 1:1,000.

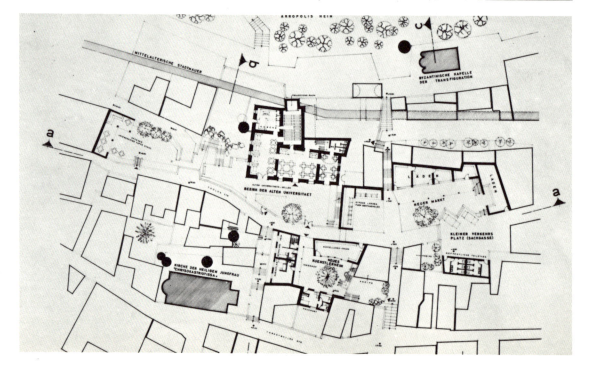

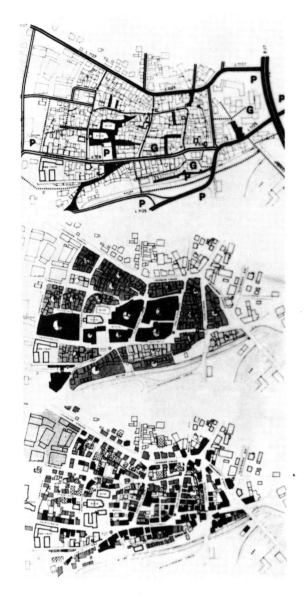

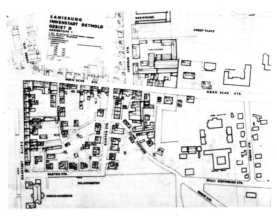

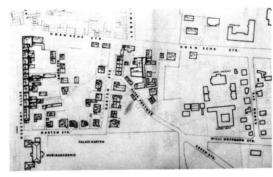

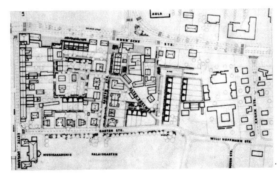

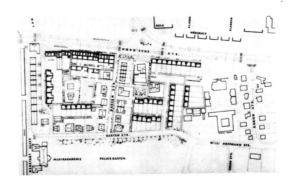

295a to 296b

Study for the radical restructuring of two residential blocks of no aesthetic interest on the outskirts of the historic centre of Detmold, Germany. This scheme provides for the demolition of over half of these blocks. Most of the sites freed in this way would be used for new buildings, which would be erected with their principal façades running parallel to the road (like the buildings they replaced). The remaining sites would be used for pedestrian walks and gardens, which would, of course, reduce the urbanization density of the area. Approximately fifty per cent of the restructured area would consist of new structures.

295a Plan of the original area showing the age and condition of the buildings, and the number of storeys they contain.

295b Plan of the area after the demolition phase.

296a Plan of the area after the first phase of the construction programme.

294a to 294c

Study for the rehabilitation of the historic centre of Bietigheim.

294a Proposed scheme for the regulation of motorized traffic: ring road providing access to the old town. Approximate scale 1:7,500.

294b Population density. Approximate scale 1:7,500.

294c Functions fulfilled by different buildings. Black areas: private houses with shops at ground level. Shaded areas: private houses. White areas: public buildings. Grey areas: industrial buildings and craft workshops. Approximate scale 1:7,500.

296b Plan of the area after the second phase of the construction programme.

A further essential requirement is the provision of an efficient method of sewage disposal. If it should prove impossible to run drains to certain houses due to the presence of valuable subterranean sites, then these houses should be provided with septic tanks. Unfortunately, the sewage systems in use in the majority of historic urban centres today are extremely primitive. Waste water is simply discharged into cesspools, which undermine the foundations of the houses and make their basements so damp that they are unfit for human habitation. Dry earth closets, which are a constant threat to public health, are still a common feature of rural urban centres.

The installation of mains, water and sewage systems in historic centres is bound to be both difficult and costly on account of the narrow roads and weak foundations. Clearly, it should be carried out before the roads are repaved.

We have already dealt at length with the question of street lighting (see 4115). But it bears repeating that if electric lights are mounted on masts and if transformer stations are given undue prominence, then the townscape of a historic centre will inevitably suffer. Electricity supply lines – like telephone lines – should be placed underground. So too should transformer stations, which could then be sited on public squares or even on inner courtyards, provided this did not interfere with freedom of access (by means of an open or enclosed passageway) between the courtyards and the urban road networks.

As for the installation of ducted urban heating networks, these would appear to be impracticable in the majority of historic centres since the roads are too narrow to accommodate the necessary pipes. The connections from the network to the individual houses would also pose major problems, for the foundations of historic houses are not very stable, which means that conversions of this kind would be inadvisable.

The preceding remarks are intended as general guidelines for the introduction of modern technical infrastructures into historic areas. The need for such installations varies considerably from centre to centre. The medieval urban sector of Saint-Germain in Paris, for example, is in a very different position as far as modern amenities are concerned from the medieval village of Mesta on the Aegean Island of Chios. Although Saint-Germain started out as a small village built around an abbey of the same name, it was destined to develop into one of the most important historic centres in the French capital. The village of Mesta, on the other hand, has always remained a small rural settlement. Because of the different social demands made on them, these two historic urban clusters have developed along completely different lines with the result that today Saint-Germain possesses a complete technical infrastructure while Mesta has none at all.

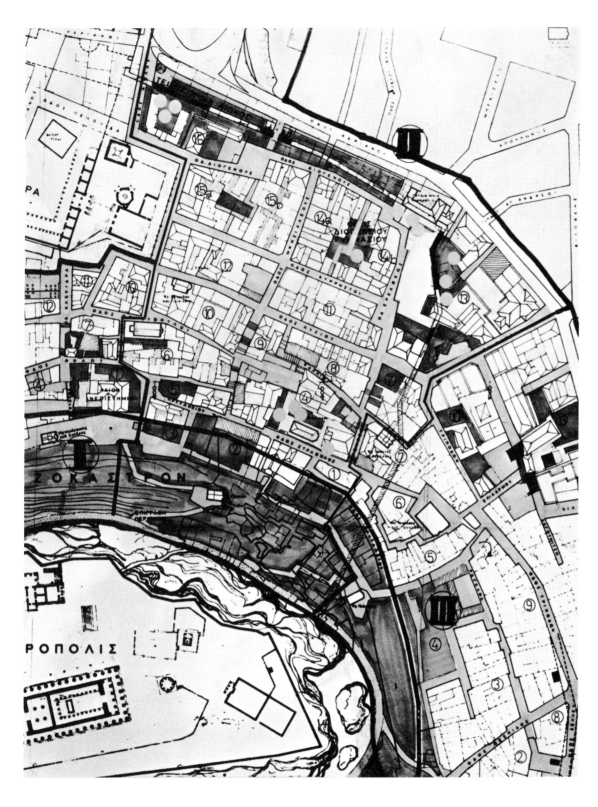

297

Plan of the scheme proposed by the author for the rehabilitation of the Plaka. Approximate scale 1:2,000. The scheme envisages the demolition of the urban area where the Roman agora is situated. The excavation of this site could then be completed and the Roman agora linked with the classical agora on the western side of the Plaka. This would remove the unsightly gaps which at present disrupt the unity of the townscape. The colony of primitive huts, which is known as the 'Anafiotika' and is situated on the northern slope of the Acropolis, where it clashes violently with this antique setting, would also be demolished. The grove surrounding the Acropolis and the antique walk at its base could then be restored. Otherwise no large scale demolition is planned for the Plaka. The historic urban cluster would be preserved and virtually all of the modest but well proportioned nineteenth-century dwellings, which provide a perfect visual transition from the modern centre of Athens to the antique monuments on top of the Acropolis, would be rehabilitated. Ten buildings at most would need to be demolished on aesthetic grounds. The road network, which is made up of narrow winding streets, would be carefully conserved and three small squares would be created so that small shopping centres could be built by converting the ground floor rooms of the houses overlooking these squares into shops. (At present there are no squares in the Plaka.) Since no point within the Plaka is more than two hundred metres from Hadrian Street, the conversion of this district into a pedestrian area will present few difficulties. A single branch road running from the main access road to the three small squares within the Plaka would be used for the delivery and collection of goods, but only in the early hours of the day. The measures proposed for the regulation of this vehicular traffic are detailed in the commentary on fig. 364.

The aesthetic significance of historic centres and the different ways in which they might be incorporated into the general spatial cluster depend on a number of factors. These include the geographical situation of the historic centre, its position in relation to later urban sectors, its communications (road, rail, etc.) with the outside world, its topography and the presence (or absence) of important architectural monuments either within its precincts or in its immediate vicinity.

In our classification of historic settlements (see 132), which was based on their importance as centres and on their geographical position, we established the following four categories:

   a) Individual and monumental groups of buildings which resemble settlements;
   b) Small rural historic centres;
   c) Historic towns;
   d) Historic urban sectors of major cities.

We also listed the most important variants within these four groups (see figs. 8a, b, c, 9).

On the basis of this classification, we now propose to investigate both actual and desirable communications networks (chiefly road routes) serving historic centres.

Each row of figs. 301, 302, 303 and 304 shows one of the main variations of four major groups of historic centres listed above (see also figs. 8a, b, c) and also illustrates the six most typical forms of communications network associated with it.

It goes without saying that these constitute only a selection from the total possible number of variants. None the less, what emerges from these illustrations is that there are five basic types of access, which appear either in isolation or in combination with one another. They are:

   1) *Access by means of branch roads terminating within the historic centre;* a major cross-country route or (in the case of historic urban sectors) an urban clearway passes close to the historic urban cluster. Access is effected by one or more branch roads terminating within the historic urban cluster.

   2) *Access by means of a through road;* a major cross-country route or an urban clearway passes through the historic urban centre.

   3) *Access by means of through branch roads;* a major cross-country route or an urban clearway passes close to the historic urban cluster. Access is effected by a branch road which passes through the historic urban cluster.

   4) *Tangential access;* a major cross-country route or an urban clearway or their branch roads touch the periphery of the historic urban centre without passing through it.

   5) *Ring road without direct access;* a major cross-country route or a branch road coming from it passes round the historic urban cluster, which is reserved for pedestrians.

The types of access illustrated in figs. 301, 302, 303 and 304 is denoted by the number or numbers in the bottom right-hand corner of each square.

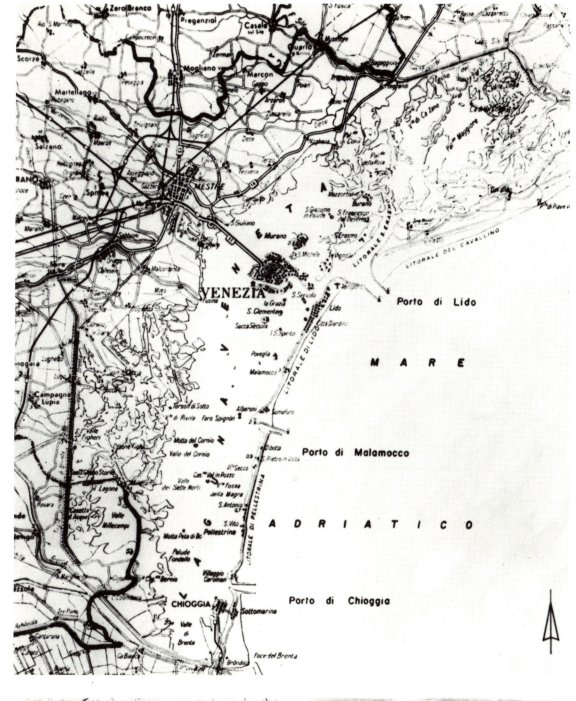

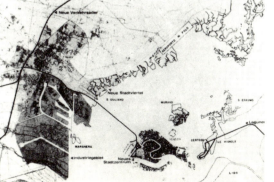

298

298

Plan of Venice. Approximate scale 1:250,000. This city, whose historic townscape has been almost completely preserved, is a perfect example of the first type of road access described on page 168. Venice was built on a group of small islands, and consequently motor cars are able to enter only the western part of the city, which is connected to the mainland by a road and rail bridge. Because of this the city has never had to contend with serious traffic problems. Ships, on the other hand, pose a grave threat, for their vibrations are undermining the foundations of Venice and the pollution of the canals has had pernicious effects on Venetian buildings.

299

Sketch showing the new developments proposed for Venice and the neighbouring mainland area. Approximate scale 1:250,000. In order to relieve the pressure on the old city, it is proposed to build a new port and a new industrial area at Maghera on the mainland (grey area). The present port, which is situated in the western part of the city, would then be converted into a new urban centre. A new road to carry tourist traffic is also planned and will run across the Lagoon to Certosa in the east. A new residential area is to be built at San Juliano on the mainland (north of the present road bridge).

300

Aerial photograph of Venice. In the foreground, the Piazza San Marco and the historic centre of the city overlooking the Canal Grande. In the background, the road and rail bridge which connects the city to the mainland with the port area on its left.

Diagrams showing the different types of communication networks associated with different types of historic centres.

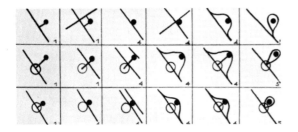

301 Networks associated with individual monumental groups of buildings.

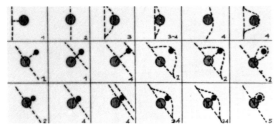

302 Networks associated with small rural historic centres.

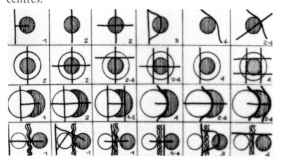

303 Networks associated with historic towns.

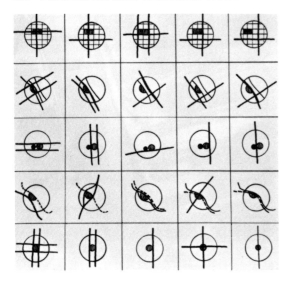

304 Networks associated with historic urban sectors of major cities.

It is hardly necessary to give a detailed description of all the different types of communications illustrated in these figures. If we describe just one of these, the remainder should be self-explanatory. The third row of fig. 302 represents a small rural historic settlement with a single group of monumental buildings (such as a monastery or a fortress) on its outskirts. Reading from left to right, the squares illustrate the following six types of road access:

a) A major traffic route passes through the historic centre (shaded circle) while a side road branches off at a point within the historic centre to run to the monumental group (small black circle). Types of access: 1 and 2.

b) A major traffic route passes between the historic centre and the group of monumental buildings, touching them both at a tangent. Type of access: 4.

c) Two major traffic routes run parallel to one another. One touches the outskirts of the historic centre and the monumental group, while the other touches the outskirts of the monumental group alone (internal and external tangents). Type of access: 4.

d) A major traffic route passes through the historic centre while a through branch road touches the outskirts of the monumental group. Types of access: 2 and 4.

e) As in (d) plus a direct link road from the historic centre to the monumental group. Types of access: 2 and 4.

f) A major traffic route passes through the historic centre while a side road branches off at a point within the centre, encircles the monumental group and returns to join the major traffic route at the same point. Types of access: 2 and 5.

Many historic centres already possess one or more of these five basic types of access road, which could easily be introduced into others. This is important, for these access roads would seem to set the pattern for the future since they guarantee the integration of historic centres into the surrounding spatial cluster.

The first type of access (by means of branch roads terminating within the historic centre) is normally found in small rural historic settlements or in historic towns situated on islands or peninsulas or in mountainous terrain. Since the major traffic route bypasses the historic urban cluster, no through traffic passes through it with the result that its urban image is unlikely to be impaired and its survival is more or less ensured. At the same time, those passing by on the major traffic route will obtain a clear view of the 'outer' townscape. In other words, the urban silhouette will be clearly recognizable in its natural setting.
Things are very different, however, in the case of the second type of access (by means of a through road).

The peace and quiet of small historic centres is completely shattered when a major traffic route passes through them. In such cases, therefore, our principal concern must be to ensure that this major route is removed from the urban cluster. This calls for a bypass – usually not more than five kilometres in length – which either touches or passes round the outskirts of the historic centre. When such bypasses are built, parking areas should be laid out for the use of visitors on the edge of the historic centre and with direct access to the bypass. This would be a particularly welcome measure since it would also tend to promote the development of pedestrian areas (see 434), no part of which need be more than five hundred metres from the bypass. Over the past twenty years, hundreds of small European provincial towns have been relieved of through traffic by means of bypass roads, which have provided them with the basis of a re-generated existence.

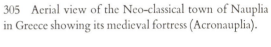

305 and 306
Examples of small historic towns with completely homogeneous urban clusters. Such towns were able to preserve their townscape because of natural or artificial barriers.

305   Aerial view of the Neo-classical town of Nauplia in Greece showing its medieval fortress (Acronauplia).

306   Aerial view of the medieval city of Nördlingen in Germany showing its circular fortifications.

307

Plan of the medieval city of Regensburg on the Danube, Germany. Approximate scale 1:30,000. Because of its decline as a regional centre (following the growth of Augsburg, Ulm and Nürnberg), Regensburg has retained its original townscape throughout the centuries. Now, however, it is threatened by progressive decay, and between 1965 and 1967 an enquiry was carried out into the feasibility of founding a new university in Regensburg and reviving the city as a regional centre. As a result, a scheme has been put forward for a new urban centre, part of which would be built above the railway yards. But this new centre would not be allowed to encroach on the historic centre. Although new ring roads would be constructed linking these two sectors, the new development would constitute a linear (and not a radial-concentric) extension of the city.

308

Plan of Mainz (Germany) in 1877. Approximate scale 1:12,500. This is an interesting example of the juxtaposition of old and new urban sectors. The Rhine in the west, the railway marshalling yards in the north and west and the fortifications in the south provide a firm framework for the medieval sector, which is thus separated from the new urban extensions of the eighteenth and nineteenth centuries.

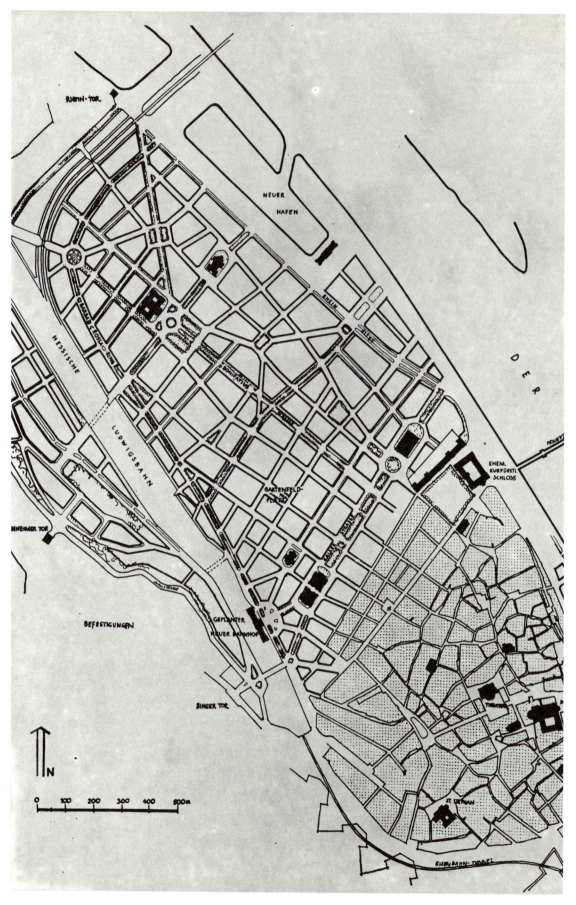

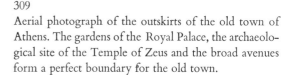

309
Aerial photograph of the outskirts of the old town of Athens. The gardens of the Royal Palace, the archaeological site of the Temple of Zeus and the broad avenues form a perfect boundary for the old town.

310
Plan of Bremen, Germany, in 1966. Approximate scale 1:14,000. The semi-elliptical moat which surrounded this Renaissance city has been replaced by a park containing various small lakes, which provide a discreet association with the past in what is otherwise a new visual transition from the old to the new sectors of the city.

311
Plan of Vienna in the middle of the nineteenth century. Here the medieval inner town with its complete ring of fortifications has been separated from the later urban sectors by a circular park. In the case of Vienna, where urban expansion followed a radial-concentric pattern, this was the best possible way of differentiating between the old and the new.

312
Aerial photograph of the Ile de la Cité in Paris. Despite its insular position, which normally facilitates the protection of an urban cluster, the historic core of Paris failed to survive the reign of Napoleon III. In order to eliminate one of the major breeding-grounds of sedition in Paris, Napoleon ordered Haussmann to demolish large numbers of houses in the densely-packed medieval centre around Notre Dame and to replace them by a new administrative centre, whose pompous edifices have completely ruined the environs of the cathedral.

In the case of large historic towns or important historic urban sectors, it would be difficult to reroute or eliminate the major roads passing through the historic cluster since they form an integral part of the urban communications network. What is more, they constitute an important element in the make-up of the townscape and, for this reason also, should be allowed to retain their present function as traffic routes for residents and visitors (see 434). At the same time, however, it is highly desirable that the volume of traffic on the roads in inner urban areas should be reduced by rerouting through traffic wherever possible. This could be done by building outer ring roads to take long distance through traffic, and inner ring roads to take urban through traffic. An excellent inner ring road system was built after the war in Munich where, together with the underground system now under construction, it helps to keep down the amount of vehicular traffic passing through the historic centre of the city. The inner road network of Florence, on the other hand, is still hopelessly congested.

The third, fourth and fifth types of access (access by means of through branch roads, tangential access and ring road without direct access) virtually preclude all possibility of through traffic entering historic centres. They also have the advantage of allowing an unimpeded view of the outer townscape. For example, a motorist driving along the *Autostrada del sole*, which bypasses Florence at a distance of some ten kilometres, obtains a perfect view of the whole city.

But urban road links with major traffic routes are not the only means by which historic centres are integrated into their spatial cluster. There are other important factors at work here.

Because of their special geographical position, which keeps down the volume of traffic in their historic sectors and tends to discourage undesirable urban developments, a number of traditional towns seem predestined to become symbolic 'points of orientation' (see 222) in the spatial clusters of the future. These towns include Amsterdam, Lübeck and Venice, which are situated on islands, and Siena, Assisi and Nauplia, which were built on high ground.

Like the monasteries of Mont-Saint-Michel and Mt. Athos, which are also situated on heights, these historic towns are cut off, both visually and functionally, from the surrounding natural and artificial cluster and, for this reason, have retained both their structural homogeneity and their symbolic power.

But in the majority of cases, there is no such clear-cut division and consequently special zones are needed which would underline the essential homogeneity of the historic urban cluster but without completely segregating it from the newer urban areas. These special zones could consist of green belts (landscaped hills and artificial lakes with old fortresses and ramparts and, in certain cases, archaeological sites [see 4114]) or, alternatively, they could be used for urban ring roads with landscaped verges.

The presence of a historic urban centre in a major city also makes an important contribution to the visual homogeneity of the total urban townscape by preserving its general scale. This is probably best illustrated by the Plaka, which covers an area of twenty-five hectares and contains modest nineteenth-century Neo-classical buildings. But, although these buildings are both aesthetically and historically far less valuable than those in other Greek Neo-classical towns (e.g. Nauplia and Syra), the Plaka none the less fulfils an extremely important function within the conurbation of Athens. Situated on the northern slope of the Acropolis, it affords an ideal visual transition from the contemporary buildings in the metropolitan area of the city (many of which are ten storeys high) to the noble architectural and natural monuments of the Acropolis, the Pnyx and the Aeropagus.

434   *The Restoration of Pedestrian Areas*

The removal of all, or nearly all, of the motorized traffic from historic centres and the creation of pedestrian areas is an essential part of preservation. By banishing the motor car, we could save the townscape from deformation and also protect historic buildings from the pernicious effects of exhaust fumes and vibration.

Of recent years, considerable thought has been given to the question of separating pedestrians from vehicular traffic in residential, cultural and commercial areas in order to reduce both air pollution and traffic accidents. This type of project – which has met with widespread approval – would also be served by the restoration of pedestrian areas. *Thus, apart from adding to the aesthetic appeal of historic centres, these pedestrian areas would also help to improve living conditions.*

We have already mentioned (see 4111) the calamitous effects produced by the unplanned introduction of vehicular traffic (trams and motor cars) into historic centres in the course of the past one hundred years. The steel bodies of the cars, the metal tram rails, the overhead electric cables, the traffic signals and lights clash violently with the traditional buildings with inevitable and disastrous repercussions on the homogeneity and aesthetic quality of the townscape.

Most of our traditional urban road networks were planned for pedestrians; their alignment, width and gradients were all worked out accordingly. Consequently, the great majority of these roads are unsuitable for mechanical transport. *And so the character of the traditional road networks also calls for the restoration of pedestrian areas.*

Unfortunately, now that administrative centres for tertiary services have been established within the historic urban sectors of our major cities, *there is no hope of totally banishing the motor car from these areas, despite the negative influence which it has undoubtedly exerted on the character of urban life.*

Given that this is so, the most expedient measure that could be taken at present to reduce the volume of surface traffic in these urban areas would be the building of new underground rail networks and underpasses linking the city centres with the outer urban areas. The response in this sphere has varied greatly. In Paris the Metro, which offers an extremely comprehensive underground service in itself, is being supplemented by the 'Métro express régional', which is now under construction. In Rome, on the other hand, there is no underground network at all.

Another measure which could be taken immediately – unfortunately a restrictive one – would be the partial or total prohibition of private transport in certain sectors of historic urban centres. This ban should be accompanied by the erection of a chain of multi-storey and underground garages on the outskirts of the sectors concerned. Although no government has had recourse to this highly unpopular measure as yet, in our opinion it is only a question of time before it is introduced on a wide scale.

In the case of small historic urban sectors, which are largely given over to residential districts and cultural centres, the more radical solution of transforming all (or most) of the road networks into pedestrian areas would appear to be perfectly feasible in certain circumstances. Its application would depend on local conditions, i.e. on the size, geographical position and structure of each individual historic urban cluster.

Contemporary town-planners have tended to make a special feature of pedestrian areas and have urged their introduction into all urban sectors, whether historic or not, which would suggest that such areas may well be still more popular in the future.

Since the end of the Second World War, there have been two distinct trends in the planning of pedestrian areas.

The first of these, which is quite radical, provides for the total segregation of all urban functions from one another. When entirely new centres are designed for tertiary services and for residential purposes (e.g. the *Défense* and *Front de Seine* projects in Paris), the urban clearways for through traffic and the regional railway lines will be sited on a number of different levels underground, the local vehicular traffic (motor cars etc.) will be carried on surface roads, while the pedestrians will be provided with raised walks at one or more levels and will be completely separated, therefore, from all traffic. Department stores and shops, cafés and public gardens will also be laid out at pedestrian level so that people will be able to meet one another and do their shopping in a completely carefree atmosphere, as in past centuries. Tall buildings of all kinds will rise up high above these pedestrian platforms and will serve as office or residential blocks. Their lower storeys will be used as garages, where shoppers, workers and residents can park their cars and where goods vehicles can load and unload. Access to the various levels will be by ramps, lifts and escalators.

This solution is the most comprehensive that we can hope for in the final phase of static town planning in which we now find ourselves but, unfortunately, it has been used on only very few occasions because of the expense.

In most redevelopment projects, especially where these involve the creation of new residential areas, a more economical system of separating pedestrians and traffic is employed. Under this system, the new urban cluster is divided up into 'communities' ranging from small neighbourhood communities to large metropolitan communities, all of which are set out at ground level and consist of freely aligned residential areas grouped around administrative, cultural, commercial or community centres. These communities, which are divided from one another by urban clearways and inner urban roads and which have parking lots on their outskirts, are planned as pedestrian precincts. Consequently, the residential blocks are linked with one another and with the different 'centres' by means of footpaths. Very few underpasses and very few underground garages are provided in such areas.

Although this kind of plan – which was a brainchild of Le Corbusier's and has been implemented on numerous occasions in Europe during the past twenty years – provides urban residents with a pedestrian area and enables them to live in direct contact with a natural setting, it has serious defects for it has produced what the French aptly refer to as the *maladie des grands ensembles*[63], i.e. a psychological condition induced in the residents of large blocks by the unaccustomed scale and layout of the architecture. When residential areas are cut off from traffic routes, road accidents are eliminated and air pollution is greatly reduced. But these gains are offset by the loss of involvement suffered by the population. The crowded

---

63 Known in England as 'Point neurosis' from the practice of naming high blocks 'Point'.

313 to 315
Examples of historic centres whose landscapes have been ruined either by the motor car or by the incorporation of modern infrastructures.

313 A street in Gamla Stan, the old sector of Stockholm, blocked by motor cars.

314 Cologne Cathedral with the monstrous nineteenth-century railway station in its immediate vicinity.

315 A historic centre in England, where rows of parked cars hide the buildings from view and completely distort the streetscape.

shopping streets of traditional urban areas gave people the feeling that they were part of the urban scene. The residents in these big new development areas have lost this sense of belonging, which has been replaced by an oppressive sense of alienation.

This kind of psychological alienation will certainly not occur in historic centres, for they possess an exceptionally coherent and dense urban structure (see 4112 and 4113) and their townscapes provide a wealth of urban experiences which promote urban activity. Moreover, if the road network of a historic urban centre is converted into a pedestrian area, this will not only protect the architectural substance of the settlement and reduce air pollution, it will also go a long way towards rehabilitating the urban image. This image or 'roadscape' is in any case not easy to observe from a vehicle. *It is essential, therefore, that we should be able to walk in peace and safety through the streets of our historic urban centres, for it is only then that we will be able to appreciate the monuments and the countless stylistic details which help to make up their urban composition.* Incidentally, shopping expeditions (to craft and souvenir shops, fashion boutiques, book and art shops) and all the other urban activities which have kept our historic centres alive, are stimulated by pedestrian traffic.

But the question that has to be answered is whether, or to what extent, it will be possible to banish the motor car from our urban road networks. Any such undertaking would certainly be accompanied by grave difficulties.

The first of these difficulties, which is primarily psychological, arises out of the growing tendency displayed by urban populations today to use mechanical and, preferably, private transport whenever possible. But it should be possible to discourage this habit – especially in small areas such as historic urban centres – by educating the public (see 425) and, failing this, by legislation, which could easily be introduced.

316

A pedestrian walk in the reconstructed centre of Hanover. The great width of this walk reminds the observer of its original function, which was to carry vehicular traffic. By arbitrarily changing a traffic route into a pedestrian walk, the planners responsible for this project have produced a nondescript urban structure lying somewhere between a road and a public square.

317

The Lijnbaan in the reconstructed centre of Rotterdam. This is a good example of a pedestrian shopping centre. It was designed as such, for the centre of Rotterdam was completely destroyed in the Second World War.

But a really major obstacle to the restoration of pedestrian areas is posed by our present-day traffic requirements. It would not be at all easy to banish motor cars from the important traffic routes which pass through the historic urban sectors of our major cities and connect them with neighbouring urban centres. Delivery and furniture vans, ambulances and fire engines must have access to these sectors.

Many attempts have been made to evolve a system which would reconcile these two diametrically opposed needs, i.e. the creation of pedestrian areas and the maintenance of traffic routes for essential services. Most of these are based on the unattainable principle of the parallel existence on the same level of two separate but interrelated networks, namely pedestrian walks and traffic routes.

In 1950 Adolf Abel suggested in his book *Die Regeneration der Städten* [The Regeneration of Cities] that sites for independent networks of pedestrian walks could be created in historic urban centres by 'gutting', i.e. removing the inner core of residential blocks. The existing road network could then be given over entirely to traffic, for the pedestrians would be able to use these inner areas. The businesses, shops and workshops now facing the roads would simply do a *volte-face* and have their main façade and display windows overlooking the pedestrian walks, which would thus become the real centre of urban activity. Goods deliveries could then be effected from the traditional road network, access to the shops being by way of the former front door.

But this solution, although highly ingenious, has two major defects.

Firstly, no matter how beautiful or interesting the new 'inner' façades were made, the original road network with its authentic façades, its squares and monuments and its perspective views must inevitably be more beautiful and more interesting. And this network would not only be exposed to the hazards of urban traffic, *it would also be concealed from the pedestrian, who would be unable to observe the traditional townscape, which is one of the greatest experiences provided by historic centres.*

Secondly, these two networks – the new pedestrian walks and the old road network – would be at the same level, which means that they would intersect one another at regular intervals of between fifty and one hundred and fifty metres. Consequently, this solution would not produce a genuine pedestrian area in which people could move about freely.

The complementary proposal made by Athanasios Aravantinos in his *Großstädtische Einkaufszentren* [Metropolitan Shopping Centres] in 1963 also suffers from the second of these two defects. Like Abel, Aravantinos recommended the progressive 'gutting' of residential blocks, although he preferred to use the new inner areas for traffic routes and to convert the traditional road networks into pedestrian areas. His scheme also has a further disadvantage for while the narrow roads of the medieval settlements would be ideal subjects for conversion, the wider roads built during the last three hundred years, which were designed for horsedrawn vehicular traffic, would not. If these were reserved for the sole use of pedestrians, they would look completely out of scale.

After the Second World War, the Germans converted numerous short but wide urban roads into shopping streets. We find examples of such conversions in the inner urban areas of most German towns. But these new pedestrian precincts are far from convincing. Their width, coupled with their rectilinear alignment, constantly remind the observer of their original function, which was to carry vehicular traffic.

In our view it is quite absurd to try to juggle with the spatial characteristics of historic urban sectors, for the problem posed by road traffic in such sectors will never be solved by proposals which envisage the coexistence of two separate networks on one and the same level. And, since it is virtually impossible to exclude road traffic completely, the only rational thing to do is to try to control it. Successful experiments along these lines have already carried out by the Swiss authorities in Zürich and Geneva. There the whole of the traditional urban area is closed to traffic for the major part of the day. But no attempt has been made to convert the actual road network. On the contrary, the pavements, the ornamental trees and the road surfaces are unchanged, which means that they have retained their original character and that the roads can still be used both for the delivery of goods (from about five to ten o'clock in the morning) and in cases of emergency.

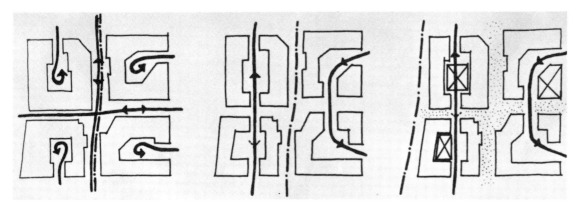

318 to 319
Sketch illustrating Adolf Abel's scheme for the creation of pedestrian areas in historic urban sectors.

318   Present condition of one such sector.

319   Condition after the creation of a pedestrian area. This operation would involve the gutting, i.e. the removal of the inner core, of large residential blocks to free sites for pedestrian walks. Shops on the ground floor of these blocks, which now face the roads, would close up their present display windows and build new ones overlooking the pedestrian area.

320, 321 and 322
Sketch illustrating the progressive gutting of four residential blocks as envisaged by Athanasios Aravantinos. In Aravantinos's scheme, the new inner areas would be used as traffic routes while most of the existing roads would be converted into pedestrian areas. The straight lines on fig. 322 indicate access roads for the delivery and collection of goods, the broken line indicates an urban traffic route, while the dotted areas represent pedestrian walks.

323
Proposed scheme for the regulation of urban traffic in the historic town of Regensburg. Approximate scale 1:10,000. In this scheme, nearly half of the existing road network and nearly all of the public squares would be given over to pedestrians. Traffic would enter the inner urban area by means of two new partial ring roads. But, although this scheme reveals an intelligent and flexible approach, it is essentially a compromise solution, for the two different networks (i.e. pedestrian and road) are to be developed on the same level, which means that they must intersect at every turning. A homogeneous pedestrian area is, therefore, out of the question.

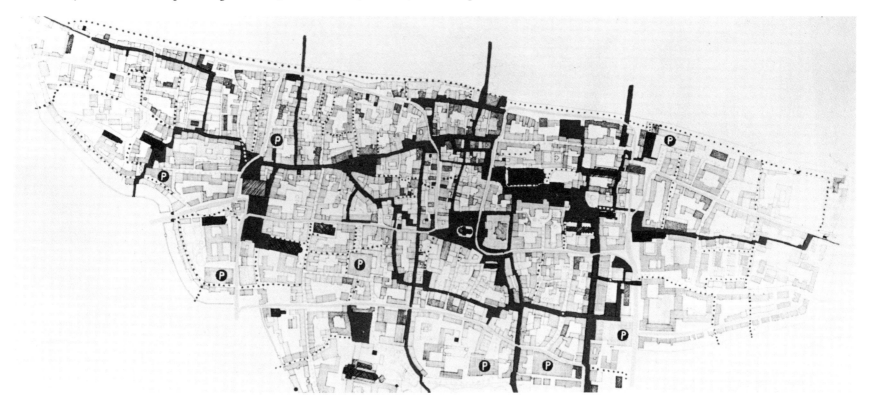

324 to 326

Two alternative schemes for the regulation of urban traffic in the centre of Munich.

324 Scheme proposed by Jensen. Approximate scale 1:25,000. This scheme envisages the creation of three different 'access areas', where vehicles moving around the outer ring road could enter the city centre. Once there, they would be allowed to use all of the side roads and alleys. By contrast, the two main traffic routes, which intersect in the city centre, would be converted into pedestrian areas. This proposal is patently absurd. To close the only roads wide enough to carry a large volume of traffic and channel this traffic into the labyrinth of sideroads and narrow alleys in the old town is to invite disaster.

326 Alternative scheme proposed by Borchert. Approximate scale 1:25,000. This is a far more sensible proposal. In Borchert's scheme, the two major traffic routes would serve as access roads for traffic entering the historic area, while many of the side roads in the area would be converted into pedestrian walks. This would help to keep the historic urban cluster free from congestion.

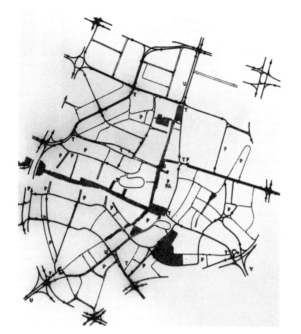

325 View of the historic urban centre of Munich. On the right the Theatinerkirche, in the centre the Feldherrenhalle.

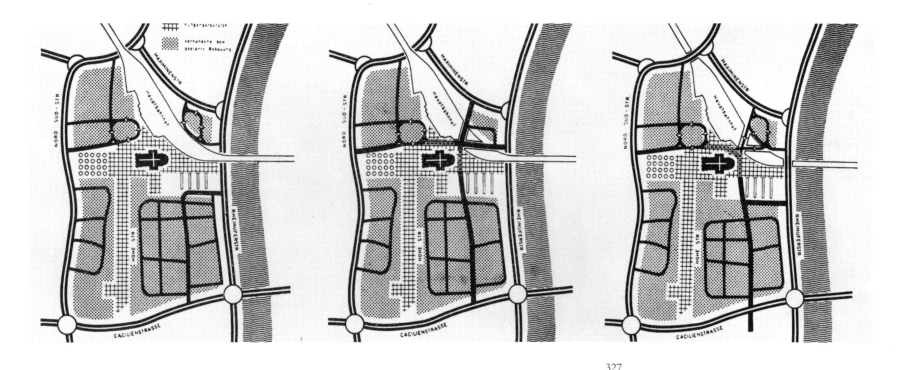

327
Three different proposals for the regulation of traffic and the creation of pedestrian walks in the environs of Cologne Cathedral. Approximate scale 1:15,000. Like the Regensburg scheme (fig. 323), these are essentially compromise solutions.

328
The Lijnbaan – the first modern pedestrian shopping centre to be built after the Second World War – is situated in the centre of Rotterdam.

329
Shopping street in Stuttgart converted into a pedestrian area.

330

Perspective sketch showing part of the Rue de Rivoli and the Jardin des Tuileries in Paris. Recently a scheme was evolved for a pedestrian area consisting of the Palais du Louvre, the forecourt of the Louvre, the Jardin des Tuileries and the arcades on the Rue de Rivoli. This would involve the construction of various underpasses to take through traffic, underground garages and parking areas, and even underground shopping centres. Once this was done, the surface would be freed for an extensive and culturally highly valuable pedestrian area.

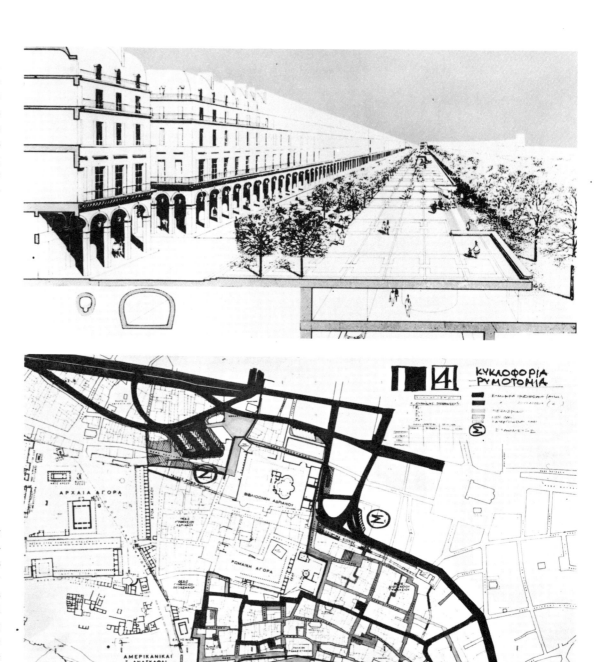

331

Plan of the Plaka, illustrating the author's proposals for the regulation of the urban traffic in the old town. Approximate scale 1:5,000. Hadrian Street and the roads giving access to the old town are marked in black. By demolishing the houses on one side of Hadrian Street, its width could be increased to twelve metres, which would enable it to carry traffic in both directions. Four parking areas (with a total capacity of three hundred cars) are planned, one at each of the four points of access to the old town. Two of these (which are marked on the plan by a capital 'S') would be underground. The four access points would be: the entrance to the integrated Roman and classical agoras (top left); the two ends of Hadrian Street (top and bottom right) and the southwestern entrance, which would also give access to the Acropolis, the Aeropagus and the classical agora (bottom left).

A branch road leading from the access road to various points in the old town would be used during the early hours of the day for the delivery and collection of goods (marked dark grey on the plan). The rest of the road network would be converted into a pedestrian area, no point of which would be more than two hundred metres from one of the parking areas (marked light grey on the plan). A few new roads and public squares would be built (shaded areas on the plan).

But in the case of larger historic urban sectors, we invariably find that the inner nucleus is traversed by one or more major roads. Since these routes serve to link the historic sector with neighbouring urban sectors, they should be retained as traffic routes. Such roads also have a further advantage in that they break down these larger sectors into two or more smaller areas, which would be more suitable as pedestrian precincts. In so far as it is possible to determine the size of such areas, an upper limit of thirty hectares would be appropriate, provided no single point was more than three hundred metres away from a traffic route or from the underground garages which would have to be built on the outskirts.

In certain cases – and provided the through traffic routes are of a suitable width – historic urban sectors might be provided with underpasses for access or for through traffic (see the proposal for the conversion of the Rue de Rivoli in Paris into a pedestrian area: fig. 330).

But in the case of historic urban clusters, where the density of urbanization is particularly high or where the stability of the historic buildings might be endangered by the construction of underpasses, the only realistic solution to the problem of road traffic is that adopted by the Swiss. By restricting vehicular entry to a few hours in the day, historic urban centres can be transformed into predominantly pedestrian areas without sustaining any damage to their townscape.

332 to 334
Proposed conversion of the environs of Notre Dame in Paris. The Roman and Early Christian layers would be exposed by excavation, and then covered by a concrete roof. This would establish an underground archaeological zone which could also be used as a museum. Access to the museum would be from a quay on the river. An underground garage is also planned. This would be situated behind the cathedral and would accommodate two hundred and fifty cars, thus removing the necessity for parking on the monumental forecourt.

332 This diagram shows the close links between the cathedral and the archaeological remains of the past two thousand years.

333 Diagram showing the proposed archaeological zone (shaded area) and the underground garage, which would be built beneath the garden at the rear of the cathedral.

334 Ground-plan showing the archaeological remains, which could be displayed in an underground museum. Access would be from the quay.

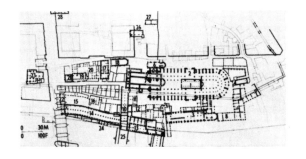

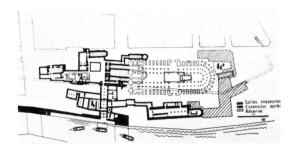

335

Sketch showing the proposed restoration of a pedestrian area in the centre of Paterson in New Jersey, U.S.A. This sketch is not unattractive and, in point of fact, it is not at all difficult to convert a nineteenth-century street into a pedestrian area 'on paper'. In practice, however, we find that the excessive width and rectilinear character of such streets makes their conversion into pedestrian areas virtually impossible. The result is almost invariably a sort of urban forum which is far too wide to be integrated into a historic centre, save on a limited scale. It would be quite absurd to imagine that a whole nineteenth-century road network could be converted into a homogeneous pedestrian precinct, since it would lack the intimacy which is such an essential part of these urban areas.

SUMMARY

Three concepts exist: Care of monuments, urban renewal, and urban rehabilitation.
The first two are subordinate to the third.
The care of monuments, which has been practised throughout the history of architecture, although not always with the same scientific precision which is its dominant feature today, is intended to preserve and protect the original substance, the form, the site and – wherever possible – the function of individual monuments or groups of monuments. Its ultimate objective is the preservation of historic townscapes.

Urban renewal, which is a much younger discipline, is intended to ensure that town-planning operations are conducted in accordance with the findings of sociological and technological research so as to create optimal living conditions (by our present standards) in old towns and urban sectors. The revival of such urban areas is achieved by the reconstruction of old buildings, the demolition of unsanitary dwellings and the provision of social amenities and technical networks (electricity, water, drainage, street lighting, etc.).
And now rehabilitation. In the ethical sphere, this word implies the restitution of the original dignity and honour of a given individual.
In the sphere of town-planning, rehabilitation means the preservation and, in some cases, the partial restoration of an original and creative conception of space and of the unique character of every important and inimitable urban cluster produced by such a conception.
Urban rehabilitation is therefore a special branch of town-planning. As such, it is a highly complex discipline and one that is concerned with a wide range of problems. But it differs from normal town-planning in so far as it deals with interventions into existing spatial compositions and not with completely new developments.

The care of monuments and urban renewal contribute to urban rehabilitation, which is in fact a far more comprehensive undertaking. By preserving individual buildings and groups of buildings and by creating adequate social and technical conditions, *those working in the field of urban rehabilitation are trying to preserve an original principle of spatial organization and to integrate it into present and future architectural creation.* Consequently, the methods and procedures employed in urban rehabilitation are more creative and more flexible than those used in the care of monuments or in urban renewal. It is possible, for example, to protect relatively unimportant buildings as 'accompanying buildings' and to remove important buildings as unaesthetic even if these are relatively new. It is also possible for entirely new buildings to take their place in preserved areas, for *the object of urban rehabilitation is not the preservation of old towns as museum pieces. On the contrary, it merely wishes to preserve the original scale of the townscape, which then has to be integrated into the new architectural cluster in its environs.*

But what is the point of all this?

It is obvious that architectural monuments should be protected and preserved because they are part of our cultural heritage.

It is equally obvious that urban sectors should be renewed because people have to live in them.

But why preserve the specific urban character of a town? Why create protected areas? For the sake of eclecticism or historicizing traditionalism? Is such a project simply a romantic illusion that is doomed to failure from the outset?

Not in our view! And this view is supported by two arguments, which have already been expounded in the main body of this book but which bear repetition.

After the ravages of the two world wars, our town-planners were tempted to introduce new development projects into traditional urban centres. But with very few exceptions – e.g. Berlin or Rotterdam, where war damage was on a gigantic scale – these proved completely impracticable. *The available urban space in these traditional centres was too limited to accommodate the new projects. Consequently, the majority of these old urban clusters now have a chance of survival.*

This then is the first, albeit negative, argument in favour of the urban rehabilitation of historic centres. The second argument is a positive one and has a bearing on the future development of town-planning.

Today we are entering upon a further phase of urban development which will be characterized by its flexibility, continuity and mobility. The new spatial megastructures and cellular structures – in which the mobile dwelling units are completely segregated from the weight-bearing frames – will be a first step in this direction.

In this new architectural age, in which urban buildings will be constantly transformed, there will be only two possible constants: nature and the static townscapes of man's first architectural age, which is now drawing to a close. These traditional townscapes – which will include those of our own 'modern' period – must be preserved so that they may be incorporated into the dynamic spatial settings of the future where they will provide a static component and bear witness to the continuity of architectural development.

To recapitulate: *Urban clusters have always had points of orientation in the past and they will probably always have them in the future. Up to now these have consisted of individual buildings or monuments. Henceforth – in the new, more comprehensive and dynamic spatial setting – they will consist of whole urban areas. Today we are no longer concerned simply with the protection of the countryside and of national customs or with the care and conservation of individual monuments. The task which now awaits us is entirely new and almost unexpected. It is to preserve the static townscape of the past so that it may be integrated into the mobile townscape of the future.*

Thus the rehabilitation of historic settlements will contribute to the diversity and beauty of the integrated spatial setting of the future.